Summer Solutions.

Minutes a Day–Mastery for a Lifetime!

Pre-Algebra

Mathematics
3rd Edition

Nancy McGraw

Simple Solutions Learning, Inc.
Beachwood, OH

Summer Solutions
Pre-Algebra
Mathematics
3rd Edition

Printed in the United States of America

ISBN: 978-1-934210-38-3

United States coin images from the United States Mint

Cover Design: Randy Reetz
Editor: Lauren Dambrogio

Instructions for Parents/Guardians

- *Summer Solutions* is an extension of the *Simple Solutions* Approach being used by thousands of children in schools across the United States.

- The 30 lessons included in each workbook are meant to review and reinforce the skills learned in the grade level just completed.

- The program is designed to be used three days per week for ten weeks to ensure retention.

- Completing the book all at one time defeats the purpose of sustained practice over the summer break.

- This book contains lesson answers in the back.

- This book also contains a "Who Knows?" drill and Help Pages that list vocabulary, solved examples, formulas, and measurement conversions.

- Lessons should be checked immediately for optimal feedback. Items that were difficult or done incorrectly should be resolved to ensure mastery.

- Adjust the use of the book to fit your summer schedule. More lessons may have to be completed during some weeks.

Summer Solutions Pre-Algebra

Reviewed Skills

- Fractions and Decimals
- Ratio, Proportion, and Percent
- Compound Probability
- Square Root and Exponents
- Integers (all operations)
- Order of Operations
- Evaluating and Simplifying Expressions
- Distributive Property and Combining Like-Terms
- One Step, Two Step, and Multi-step Equations
- Algebraic Phrases
- Equations with Variables on Both Sides
- Graphing on a Coordinate Plane
- Greatest Common Factor and Least Common Multiple of Algebraic Expressions
- Geometry Formulas
- Word Problems

Help Pages begin on page 63.

Answers to Lessons begin on page 87.

Lesson #1

1. $\frac{7}{8} \times \frac{16}{28} = ?$

2. What is the P(1, 6, 4) on 3 rolls of a die?

3. Write the number twelve and three hundred fifty-seven thousandths as a decimal.

4. $4(4a - 2b + 5) + 2(2a - 5) = ?$

5. $9 - 6\frac{5}{6} = ?$

6. $81 \div 9 + 2(5 - 2) + 3 = ?$

7. $-86 - (-37) = ?$

8. $\frac{-360}{5} = ?$

9. Find 60% of 200.

10. $-12 \bigcirc -46$

11. How many quarts are 16 gallons?

12. Solve for x. $\frac{x}{7} = 13$

13. What value of a makes the equation true? $9a + 2 = 4a - 18$

14. What is the value of x? $x - 37 = -66$

15. Write 18% as a decimal and as a reduced fraction.

16. $7{,}035 - 2{,}653 = ?$

17. Find the surface area of the rectangular prism.

18. Solve for x. $3x + 5 = -10$

19. When these fractions are equivalent, what is the value of x? $\frac{6}{12} = \frac{x}{72}$

20. $1\frac{2}{3} \div 3 = ?$

1.	2.	3.	4.
5.	6.	7.	8.
9.	10.	11.	12.
13.	14.	15.	16.
17.	18.	19.	20.

Lesson #2

1.   Solve for x. $2x - 4 = 24$

2.   What is the area of the triangle?

3. $7 + 2[4 + 2(3 + 2)] = ?$

4. $77 + (-25) = ?$

5. $14\frac{1}{3} + 12\frac{2}{5} = ?$

6.   Find the value of x. $\frac{1}{7}x = 14$

7.   Find the missing measurement.

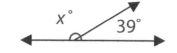

8.   Solve for a. $a + 18 = -43$

9.   Write $\frac{7}{50}$ as a decimal and as a percent.

10. What value of x makes the equation true? $5x + 8 = 7x$

11. When $x = 5$ and $y = 3$, what is the value of $\frac{xy}{5} + 3$?

12. Write an algebraic expression to represent eight minus six times a number.

13. How many feet are in 4 miles?

14. Solve for a. $\frac{1}{8}a + 4 = 12$

15. $3(-12) = ?$

16. Draw 2 congruent hexagons.

17. $39.45 + 7.552 = ?$

18. Put these measurements in increasing order.
     quart   cup   pint   gallon

19. Write the ratio 5:8 in two other ways.

20. Mr. Lamond bought a new engine for his antique truck. The engine cost $2,500, and the sales tax rate was 8%. What was the total cost of the engine?

1.	2.	3.	4.
5.	6.	7.	8.
9.	10.	11.	12.
13.	14.	15.	16.
17.	18.	19.	20.

Lesson #3

1. Solve for a. $\frac{1}{4}a - 7 = -12$

2. On a trip from Fort Wayne to Des Moines, Mr. Jackson traveled an average of 68 miles per hour. If his trip took 7 hours, how many miles did he drive?

3. $11 - 5\frac{2}{5} = ?$

4. What is the value of x? $x + 51 = 132$

5. Write 85% as a decimal and as a reduced fraction.

6. Determine the area of a parallelogram if its base is 14 cm and its height is 6 cm.

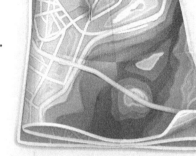

7. $93 - (-34) = ?$

8. $3(2x + 5y - 8) + 2(2x + 3y - 4) = ?$

9. Figures with the same size and shape are _____.

10. Calculate the area of this trapezoid.

11. $(5 \cdot 3 + 4 - 1) \div 6 = ?$

12. $3.4 \times 0.3 = ?$

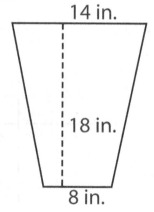

13. Find the GCF of $10a^6b^2c$ and $15a^3b^3c$.

14. If $x = 2$ and $y = 4$, what is the value of $4(3x + 3y)$?

15. What is the probability of getting a six with one roll of a die?

16. Solve for x. $3x + 7 = -26$

17. Twice a number plus 4 is twelve. Express this sentence in an equation.

18. What percent of 90 is 72?

19. Find the value of x. $5x = -45$

20. Two angles whose measures add up to 90° are _____ angles.

1.	2.	3.	4.
5.	6.	7.	8.
9.	10.	11.	12.
13.	14.	15.	16.
17.	18.	19.	20.

Lesson #4

1. $-55 + (26) = ?$

2. $0.3 - 0.144 = ?$

3.   How many centimeters are in 9 meters?

4.   A triangle with no sides congruent is _____.

5. $33\frac{1}{9} - 14\frac{8}{9} = ?$

6. $48.04 \div 0.04 = ?$

7.   Find the GCF of 12 and 18.

8.   Solve for a. $a - 22 = 61$

9.   What is the average of 24, 15, 27, 34, and 10?

10. $\frac{8}{10} \div \frac{1}{4} = ?$

11.   Solve for x. $3x + 3 = -15$

12. $29 - (-16) = ?$

13.   What value of a makes the equation true? $a + 15 = 4a$

14.   Joseph was 5 feet 4 inches tall. How many inches tall was Joseph?

15.   Solve for x. $\frac{x}{9} = 12$

16.   Find the circumference of the circle.

17.   Find the value of $(5a - 2b) \div 2$ when $a = 2$ and $b = 3$.

18.   Round 56,845,003 to the nearest ten million.

19. $0.005 \times 0.007 = ?$

20.   How many cups are in 8 pints?

1.	2.	3.	4.
5.	6.	7.	8.
9.	10.	11.	12.
13.	14.	15.	16.
17.	18.	19.	20.

Lesson #5

1.   Solve to find the value of y. $3y - 12 = 3$

2. $\sqrt{81} + \sqrt{49} = ?$

3. $64 \div 8 + 3 \cdot 4 - 2 = ?$

4.   Translate into an algebraic expression:
     Seven times a number decreased by ten.

5.   Solve for a. $6a + 5 = 41$

6. $66 + (-31) = ?$

7.   Give the perimeter and the area of a
     square with sides that measure 9 cm.

8. $\frac{-64}{-4} = ?$

9.   Factor. $45x^2y^2$

10. $\frac{5}{9} \times \frac{12}{15} = ?$

11. $1\frac{1}{2} \times \frac{2}{3} = ?$

12. $70,000 - 42,817 = ?$

13. What will be the time 90 minutes after noon?

14. $3\frac{1}{4} + 6\frac{2}{3} = ?$

15. Put these integers in increasing order. $-44, -16, -61, -2$

16. Solve for x. $3(x + 7) = 27$

17. What value of x will make the fractions equivalent? $\frac{8}{12} = \frac{x}{72}$

18. Write 13.001 using words.

19. On the Fahrenheit scale, water boils at _____.

20. Find the measure of the angle that is complementary
     to the angle shown.

37°

1.	2.	3.	4.
5.	6.	7.	8.
9.	10.	11.	12.
13.	14.	15.	16.
17.	18.	19.	20.

Lesson #6

1. Round 46.278 to the nearest tenth.

2. Solve for x. $-2(x+3)=6$

3. Find the value of a. $-7a = 84$

4. If Maurice got 12 out of 15 problems correct on his social studies test, what percent of the problems did he get correct?

5. $-95 - (-47) = ?$

6. How many gallons are 12 quarts?

7. $\frac{9}{10} \div \frac{3}{5} = ?$

8. Solve for a. $\frac{1}{3}a - 7 = 19$

9. $3.645 \div 5 = ?$

10. Draw intersecting lines.

11. $-38 \bigcirc -9$

12. Rewrite a number divided by six using algebraic symbols.

13. Find $\frac{5}{7}$ of 35.

14. $4[5 + 2(3) - 3] = ?$

15. What value of a makes the equation true? $a + 21 = -56$

16. Find the LCM of $12ab^3c^2$ and $16a^3b^4c$.

17. Write $\frac{3}{20}$ as a decimal and as a percent.

18. $4,568 \times 6 = ?$

19. Calculate the area of the trapezoid.

20. A decagon has how many sides?

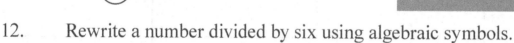

1.	2.	3.	4.
5.	6.	7.	8.
9.	10.	11.	12.
13.	14.	15.	16.
17.	18.	19.	20.

Lesson #7

1. $67 + (-35) = ?$

2.   Round 954,766,322 to the nearest hundred million.

3. $-7(-3)(-3) = ?$

4. $12\frac{1}{6} + 5\frac{2}{3} = ?$

5.   At the store, there were 6 bags of chips for every 8 cookie boxes. If there were 104 cookie boxes, how many bags of chips were there?

6.   Solve to find the value of x. $2x + 4x - 6 = 12$

7.   Find the measure of the missing angle.

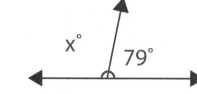

8. $33.5 + 196.77 = ?$

9.   Simplify. $9(3a + 2b + 3c - 5)$

10.   Evaluate $a(b + 7)$ when $a = 2$ and $b = 4$.

11.   Write $\frac{1}{8}$ as a decimal.

12.   What is the value of x? $\frac{1}{5}x = 22$

13.   Solve for a. $2a + 6 = -a - 6$

14.   How many yards are in a mile?

15.   Solve for x. $\frac{x}{9} = 21$

16.   It snowed 5.312 cm on Saturday and 4.24 cm on Sunday. How much snow fell during the two days? (Round to the tenth of a centimeter.)

17.   Put these decimals in order from greatest to least.

            2.001   2.2   2.12   2.02

18.   Find the area of a circle with a radius of 7 cm.

19.   What number is 80% of 200?

20. $82 - (-67) = ?$

1.	2.	3.	4.
5.	6.	7.	8.
9.	10.	11.	12.
13.	14.	15.	16.
17.	18.	19.	20.

Lesson #8

1. Find the GCF of $12xy^4z^3$ and $18x^3y^2z^5$.

2. $95 + (-58) = ?$

3. $\frac{-384}{2} = ?$

4. How many pints are in 7 quarts?

5. Solve for a. $6a - 2a - 4 = -60$

6. Translate into an algebraic expression: The product of three and a number.

7. Draw parallel, horizontal lines.

8. Find the value of a. $a - 6 = 32$

9. $\sqrt{64} - \sqrt{25} = ?$

10. $16 - 4\frac{3}{7} = ?$

11. What is the value of x? $\frac{x}{10} = -14$

12. Angles whose measures add up to 180° are _____ angles.

13. Find $\frac{4}{7}$ of 63.

14. $8 + 3[5 + 2(12 - 8) - 1] = ?$

15. Find the volume of the rectangular prism.

8 ft
5 ft
2 ft

16. $0.008 \times 0.008 = ?$

17. Put $\frac{9}{27}$ in simplest form.

18. What value of x makes the equation true? $5x - 9 = 26$

19. Write the ratio $\frac{3}{4}$ in two other ways.

20. Solve for x. $\frac{7}{10} = \frac{x}{50}$

1.	2.	3.	4.
5.	6.	7.	8.
9.	10.	11.	12.
13.	14.	15.	16.
17.	18.	19.	20.

Lesson #9

1. If $a = 5$ and $b = 3$, what is the value of $3a - 2 + b$?

2. Solve the equation to find the value of x. $4(x + 2) = 12$

3. $-54 - (-12) = ?$

4. What is the area of the triangle shown?

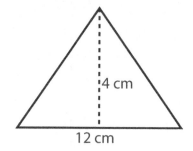

4 cm

12 cm

5. Solve for b. $b - 32 = 76$

6. How many ounces are in 6 pounds?

7. $\frac{-650}{-5} = ?$

8. Find the value of x. $\frac{x}{11} = 13$

9. Simplify. $3(6x + 4y - 7z - 4)$

10. Solve for x. $5x + 7 = -13$

11. What is the P(1, 4) on 2 rolls of a die?

12. On the Celsius scale, water boils at _____.

13. Find the value of 5^4.

14. $40 - 5 \cdot 2 + 8 \div 2 - 1 = ?$

15. Factor. $20a^2b$

16. Solve for b. $7b = -105$

17. Write 65% as a decimal and as a reduced fraction.

18. Figures with the same shape, but different sizes are _____.

19. Round 10.5723 to the nearest hundredth.

20. Use the protractor to find the measure of each angle.

    $\angle AOD =$ ____

    $\angle GOE =$ ____

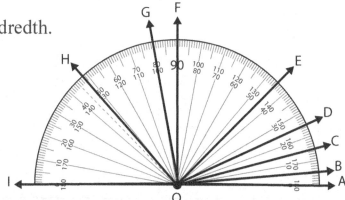

1.	2.	3.	4.
5.	6.	7.	8.
9.	10.	11.	12.
13.	14.	15.	16.
17.	18.	19.	20.

Lesson #10

1. $83 - (-30) = ?$

2. Write an algebraic phrase to express the following: The sum of a number and twelve.

3. Solve for x. $2x + 10 = 18$

4. How many pounds are 4 tons?

5. Find the area of the parallelogram.

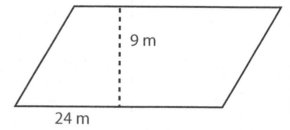

9 m

24 m

6. $6{,}125 - 2{,}978 = ?$

7. $-7(-7) = ?$

8. $6 \cdot 4 - 10 \div 2 + 3(2) = ?$

9. Find the LCM of $8a^4b^2$ and $15a^3b$.

10. Solve for x. $\frac{x}{9} = -15$

11. $-|-24| = ?$

12. What is the value of b? $-8b = -96$

13. How many years are 5 centuries?

14. $-41 + (-18) = ?$

15. Find 60% of 25.

16. The answer to a subtraction problem is the _____.

17. Which is greater, $\frac{3}{5}$ or 0.53?

18. Write 0.32 as a reduced fraction and as a percent.

19. What value of x makes the equation true? $5x - 4 = 3x - 24$

20. The picnic breakfast cost $12. Jasmine left a tip that was 15 percent of the cost of the meal. How much money was the tip that Jasmine left?

1.	2.	3.	4.
5.	6.	7.	8.
9.	10.	11.	12.
13.	14.	15.	16.
17.	18.	19.	20.

Lesson #11

1. Solve for x. $\frac{x}{3} + 7 = 31$

2. $13\frac{2}{7} - 8\frac{6}{7} = ?$

3. How many feet are in 9 yards?

4. Identify the figure by name.

5. Solve for a. $\frac{1}{9}a + 3 = -20$

6. $-58 \bigcirc -23$

7. $\frac{7}{9} \times \frac{12}{14} = ?$

8. What is the value of x? $-5x = -65$

9. $0.2 - 0.117 = ?$

10. The ratio of bananas to oranges at the market was 2 to 5. If there were 150 bananas at the market, how many oranges were there?

11. $64 - (-38) = ?$

12. The answer to a multiplication problem is the _____.

13. $3.4 \div 0.5 = ?$

14. Write an algebraic expression that means seven divided by a number, decreased by three.

15. $9,333 \times 4 = ?$

16. How many degrees are in a straight angle?

17. $\frac{5}{6} \div \frac{2}{3} = ?$

18. What is the value of a in the equation? $a + 67 = -100$

19. Solve for a. $4a + 8 = 6a - 4$

20. A triangle with all sides congruent is a(n) _____ triangle.

1.	2.	3.	4.
5.	6.	7.	8.
9.	10.	11.	12.
13.	14.	15.	16.
17.	18.	19.	20.

Lesson #12

1. Find the circumference of a circle if its diameter is 14 inches.

2. $0.47 - 0.2156 = ?$

3. How many tons are 12,000 pounds?

4. What value of x makes the equation true? $3x + 8 = 5x - 6$

5. $6\frac{1}{8} + 6\frac{1}{2} = ?$

6. Solve for a. $a + 51 = -83$

7. Of the 80 horses in the field, five-eighths were Thoroughbreds. The rest were Quarter Horses. How many horses in the field were Quarter Horses?

8. $-73 + (-48) = ?$

9. Solve for x. $4x - 8 = 20$

10. Find the GCF of $12a^4b^2$ and $18a^2b$.

11. When $x = 6$ and $y = 2$, what is the value of $\frac{xy}{4} + 8$?

12. Write $\frac{3}{50}$ as a decimal and as a percent.

13. Find the value of x. $\frac{1}{8}x = 11$

14. Find the missing measurement.

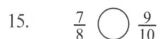

15. $\frac{7}{8} \bigcirc \frac{9}{10}$

16. What percent of 60 is 18?

17. $9.5 \div 0.05 = ?$

18. Draw a pentagon.

19. A triangle with 2 sides congruent is a(n) _____ triangle.

20. Solve for x. $-7x = 147$

1.	2.	3.	4.
5.	6.	7.	8.
9.	10.	11.	12.
13.	14.	15.	16.
17.	18.	19.	20.

Lesson #13

1. Simplify. $4(3a - 5b + 2) - 3(2a + 2)$

2. $\sqrt{36} + \sqrt{16} - 2^2 = ?$

3. $40 - 5 \cdot 6 + 12 \div 4 - 1 = ?$

4. $95 - (-10) = ?$

5. Determine the value of $(8x + y)$ when $x = 2$ and $y = 4$.

6. $\frac{5}{9} \times \frac{6}{15} = ?$

7. $-5(6)(-2) = ?$

8. How many cups are in 6 pints?

9. Which is greater, 0.56 or 75%?

10. $902,417 - 567,388 = ?$

11. Solve for a. $a + 17 = -55$

12. Twelve times a number divided by two. Express this phrase using algebraic symbols.

13. Round 95.6345 to the nearest hundredth.

14. Solve for x. $\frac{3}{8} = \frac{x}{136}$

15. $62 \times 43 = ?$

16. $-32 \bigcirc -28$

17. What is the value of x? $2(8 + x) = 22$

18. Solve for x. $\frac{x}{6} = -16$

19. Write the first five prime numbers.

20. What value of x makes the equation true? $5x - 3 = 2x + 12$

1.	2.	3.	4.
5.	6.	7.	8.
9.	10.	11.	12.
13.	14.	15.	16.
17.	18.	19.	20.

Lesson #14

1. The average weight of the three bear cubs at the wildlife preserve is 288 pounds. Two of the cubs weigh 294 and 268 pounds. What is the weight of the third cub?

2. $0.009 \times 0.08 = ?$

3. Solve for a. $a + 25 = -99$

4. How many quarts are in 15 gallons?

5. Find $\frac{6}{7}$ of 63.

6. $-22 - (-22) = ?$

7. Solve the equation for x. $\frac{1}{8}x + 4 = 18$

8. Find the area of a parallelogram if its base is 17 m and its height is 8 m.

9. $|18| = ?$

10. Solve for x. $4(x + 1) = 2x - 2$

11. Find the value of a. $-9a = 234$

12. Write 44% as a decimal and as a reduced fraction.

13. $16 - 7\frac{3}{4} = ?$

14. $\frac{-64}{2} = ?$

15. Find 35% of 40.

16. Solve for b. $b - 31 = 73$

17. Round 753,323,169 to the nearest ten million.

18. Write the formula for finding the area of a triangle.

19. Factor $30a^2b^3c$.

20. Solve for x. $\frac{x}{14} = 25$

1.	2.	3.	4.
5.	6.	7.	8.
9.	10.	11.	12.
13.	14.	15.	16.
17.	18.	19.	20.

Lesson #15

1. The number that occurs most often in a set of numbers is the _____.

2. $5\frac{4}{5} + 7\frac{2}{3} = ?$

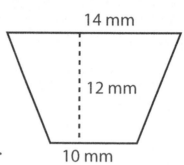
14 mm
12 mm
10 mm

3. Solve for a. $\frac{1}{5}a - 3 = 13$

4. What is the value of x? $\frac{x}{25} = -4$

5. Calculate the area of the trapezoid.

6. If $a = 4$ and $b = 3$, what is the value of $7a + 3b - 5$?

7. Find the LCM of $8x^4y^3z^2$ and $14x^5y^2z$.

8. How many ounces are in 7 pounds?

9. What value of x makes the equation true? $6a - 2 = a + 13$

10. $8.652 \div 0.2 = ?$

11. $42 - (-9) = ?$

12. Draw 2 similar hexagons.

13. $10 + 3[5 + 10 \div 2] + 4 = ?$

14. Solve for x. $12x = 180$

15. Write as an algebraic expression: The quotient of a number and five increased by nine.

16. $3a + 7 = 5a - 7$

17. Write $4 \times 4 \times 4 \times 4 \times 4$ using a base and an exponent.

18. What is the P(H, H, H, T, T) on five flips of a coin?

19. $934,259,663 + 837,755,377 = ?$

20. Which is greater, $\frac{11}{20}$ or 50%?

1.	2.	3.	4.
5.	6.	7.	8.
9.	10.	11.	12.
13.	14.	15.	16.
17.	18.	19.	20.

Lesson #16

1.   Solve for b. $5b + 3 = 9b - 1$

2. $-|87| = ?$

3. $57 - (-21) = ?$

4. $6\frac{1}{5} - 2\frac{4}{5} = ?$

5.   Factor. $18x^2y$

6. $\frac{5}{7} \times \frac{14}{15} = ?$

7.   How many centuries are 500 years?

8.   What is the value of a? $2a + 42 = -26$

9.   Evaluate $7x + 3y - 10$ when $x = 3$ and $y = 2$.

10.   Find the median of 17, 56, 37, 40, and 78.

11. $65.642 + 9.3 = ?$

12.   If a baby weighs 120 ounces, what is the baby's weight in pounds and ounces?

13.   Solve for a. $-7a = 133$

14. $8{,}502{,}301 - 3{,}686{,}895 = ?$

15. $-134 + (-86) = ?$

16.   What is the value of b in the equation? $b - 38 = -79$

17.   Find the GCF of 14 and 21.

18.   Write 0.38 as a percent and as a reduced fraction.

19.   The area of the square is 121 ft². What is the measure of each side?

20.   A drawing has a scale where 1 inch is equal to 8 feet. What is the distance on the drawing if the actual distance is 24 feet? (Hint: Use a proportion to solve the problem.)

1.	2.	3.	4.
5.	6.	7.	8.
9.	10.	11.	12.
13.	14.	15.	16.
17.	18.	19.	20.

Lesson #17

1. Write $\frac{7}{25}$ as a decimal.

2. $2\frac{5}{8} + 9\frac{1}{4} = ?$

3. Solve for a. $a + 31 = 56$

4. Find $\frac{8}{9}$ of 54.

5. $\frac{-366}{-6} = ?$

6. $0.005 \times 0.007 = ?$

7. $-50 \bigcirc -8$

8. Of the 50 students in the Art Club, 14 are boys. What percent of the students are boys?

9. Write an algebraic expression for eleven decreased by a number.

10. A triangle with no congruent sides is a(n) _____ triangle.

11. Which digit is in the thousandths place in 5.461?

12. $66 + (-47) = ?$

13. Simplify. $2(7x - 4y + 2) - 3(2x + 2y - 2)$

14. Find the area of the circle.

15. $\frac{9}{10} \div \frac{3}{5} = ?$

    6 m

16. Factor. $20x^2 y^3 z$

17. Solve for x. $16x = 96$

18. What value of x makes the equation true? $7x = 80 + 9x$

19. $50 - 5 \cdot 5 + 12 \div 3 - 4 = ?$

20. Solve for x. $\frac{x}{4} + 12 = 25$

1.	2.	3.	4.
5.	6.	7.	8.
9.	10.	11.	12.
13.	14.	15.	16.
17.	18.	19.	20.

Lesson #18

1. What value of x will make the fractions equivalent? $\frac{6}{x} = \frac{3}{7}$

2. Solve for c. $c - 85 = 109$

3. Find the GCF of $15x^2y^2z$ and $25xyz^2$.

4. $709.8 \div 0.6 = ?$

5. Find the value of a. $3a - 7 = 14$

6. Put these integers in increasing order.

    $18 \qquad -12 \qquad 0 \qquad -25$

7. If $a = 3$ and $b = 4$, what is the value of $ab + ab$?

8. Find the LCM of $12a^3b^2$ and $20a^4b^2$.

9. Solve for x. $\frac{1}{7}x + 6 = 12$

10. Write an algebraic expression for six times a number decreased by eight.

11. $\sqrt{256} = ?$

12. Round 42.3184 to the nearest hundredth.

13. $\frac{6}{7} \bigcirc \frac{8}{9}$

14. $-53 + (-28) + 11 = ?$

15. What percent of 50 is 15?

16. Find the area of the trapezoid.

17. Solve for x. $\frac{6}{x} = \frac{54}{63}$

18. $-|-34| = ?$

19. What is the value of x? $x - 3.3 = 6.2$

20. Write the coordinates of each point.

    F _____   G _____   H _____

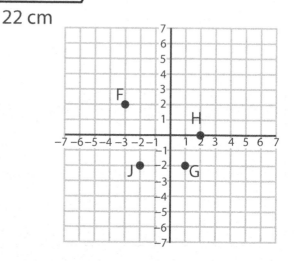

1.

2.

3.

4.

5.

6.

7.

8.

9.

10.

11.

12.

13.

14.

15.

16.

17.

18.

19.

20.

Lesson #19

1. $\frac{8}{9} \times \frac{18}{24} = ?$

2. $7 \cdot 9 \div 3 + 5 = ?$

3.   How many degrees are in a straight angle?

4.   Find the value of 3^3.

5.   What is the P(6, 1, 2) on 3 rolls of a die?

6. $0.456 \div 6 = ?$

7.   Write $\frac{4}{25}$ as a decimal and as a percent.

8. $10 - 3\frac{5}{6} = ?$

9.   How many yards are in a mile?

10.   Find $\frac{3}{7}$ of 42.

11.   Solve for x. $x - 44 = -81$

12.   Factor. $14a^3b^2$

13.   What is the value of m in the equation? $4m + 5 = 9m$

14.   When $x = 4$ and $y = 6$, evaluate $xy - x$.

15.   Write the formula for finding the area of a parallelogram.

16.   What is the boiling point of water on the Celsius scale?

17.   Solve for b. $5b = -85$

18.   Find the GCF of $10a^3b^2$ and $12ab$.

19. $\frac{3}{5} + \frac{1}{2} = ?$

20.   Which points have the given coordinates?

      a) (3, 2)   b) (2, 3)

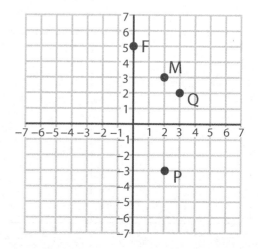

1.	2.	3.	4.
5.	6.	7.	8.
9.	10.	11.	12.
13.	14.	15.	16.
17.	18.	19.	20.

Lesson #20

1. Find the GCF of $12a^3b^4c^2$ and $15ab^2c$.

2. Solve the equation to find the value of x. $9x + 3 = -15$

3. Simplify. $5(2x + 2y + 5) + 4(2x - 2y - 6)$

4. Solve for y. $y + 19 = -76$

5. $0.7 - 0.448 = ?$

6. Solve for x. $-2x = 56$

7. Make a factor tree for 90.

8. What value of a makes the equation true? $5a + 9 = 2a$

9. List the first 6 prime numbers.

10. Three-fifths of the 60 pieces of gum in a bag were red. How many pieces of gum were red?

11. Solve for x. $\frac{x}{13} = -15$

12. Which is greater, 0.39 or 35%?

13. $7 + 2[5 + 10 \div 2 - 1] = ?$

14. Find the LCM of $6x^2y^3z$ and $9xy^2z^2$.

15. What is the value of x? $\frac{1}{10}x - 6 = 15$

16. Draw perpendicular lines.

17. Write the decimal 0.9315 using words.

18. Find the value of x. $-11 = x - 8$

19. $\frac{4}{5} \times \frac{10}{12} = ?$

20. Find the volume of the solid figure.

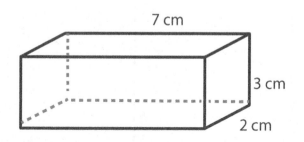

1.	2.	3.	4.
5.	6.	7.	8.
9.	10.	11.	12.
13.	14.	15.	16.
17.	18.	19.	20.

Lesson #21

1.   Write the formula for finding the circumference of a circle.

2.   Simplify. $5(4a + 3b - 2) - 3(2a - 4b + 3)$

3.   Draw 2 congruent pentagons.

4.   Find the average of 15, 18, and 21.

5.   Find the value of a. $5a + 6 = -24$

6. $64 \div 8 \cdot 4 + 8 - 1 = ?$

7.   The area of a rectangle is 150 square centimeters. If the length of the rectangle is 10 cm, what is its width?

8.   How many teaspoons are in 5 tablespoons?

9.   Monica made a drawing of her patio using the scale of 1 inch equals 4 feet. The length of the patio in her drawing was 9 inches. What was the length of the actual patio?

10. Write $\frac{4}{5}$ as a decimal and as a percent.

11. Two times a number divided by seven is forty. Express this sentence using algebraic symbols.

12. Find the GCF of $10a^4b^2c$ and $18a^2bc^2$.

13. What value of x makes the equation true? $3x + 4 = x + 18$

14. $\frac{224}{-4} = ?$

15. $\sqrt{169} = ?$

16. What percent of 30 is 27?

17. $-83 - (-44) + 13 = ?$

18. Solve for b. $b - 22 = -76$

19. Solve for x. $\frac{x}{4} - 8 = 12$

20. $7{,}000 - 2{,}956 = ?$

1.	2.	3.	4.
5.	6.	7.	8.
9.	10.	11.	12.
13.	14.	15.	16.
17.	18.	19.	20.

Lesson #22

1.   Solve for x. $x + 19 = 36$

2.   Find the LCM of $8x^3y^4z^2$ and $12x^2y^3z$.

3.   Evaluate $(xy + x) - 7$ when $x = 6$ and $y = 5$.

4.   Calculate the area of this trapezoid.

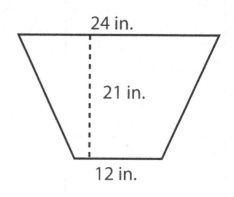

5. $|67| = ?$

6.   What is the value of x? $\frac{1}{8}x - 14 = 12$

7. $6 + 7[15 \div 3 + 2 \cdot 3] = ?$

8.   Find the missing measurement.

9.   How many feet are in 5 miles?

10.   Solve for b. $7b + 12 = -51$

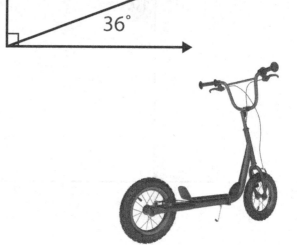

11.   Which is greater, $\frac{9}{20}$ or 40%?

12.   Factor. $10a^3b$

13. $67.7 + 8.514 = ?$

14. $9\frac{2}{9} - 3\frac{5}{9} = ?$

15.   Find $\frac{5}{6}$ of 24.

16.   Solve the proportion to find x. $\frac{9}{12} = \frac{x}{144}$

17.   Solve for b. $3b - 23 = 25$

18.   Which day had the greatest decrease in temperature?

19.   What is the average temperature for these 5 days?

20.   What was the range in the daily high temperature during this week?

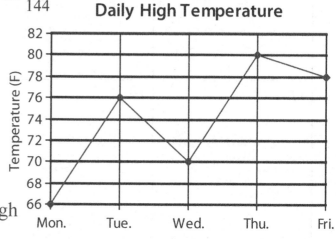

Daily High Temperature

1.	2.	3.	4.
5.	6.	7.	8.
9.	10.	11.	12.
13.	14.	15.	16.
17.	18.	19.	20.

Lesson #23

1. $-85 + (-48) = ?$

2. $\sqrt{64} + \sqrt{25} + 2^2 = ?$

3. $-44 \bigcirc -2$

4. How many cups are in 8 pints?

5. What is the value of a? $2a - 5 = 8a + 7$

6. Find the value of $3x - 4y$ when $x = 6$ and $y = 3$.

7. On the Fahrenheit scale, water boils at _____.

8. $0.62 \div 0.4 = ?$

9. Find the GCF of $12x^4y^2z$ and $16x^3yz^2$.

10. What is the area of a triangle with a base of 21 m and a height of 6 m?

11. Three students, in a class of twenty, scored a 100% on their English test. What percent of the students scored 100%?

12. $\frac{8}{9} \times \frac{27}{32} = ?$

13. Write an algebraic expression for a number decreased by fourteen.

14. Solve for x. $x + 9.37 = 12.4$

15. $6 - 2\frac{4}{5} = ?$

16. Solve for b. $b - 26 = 69$

17. Factor. $25a^2bc$

18. Find the value of x. $\frac{1}{10}x - 6 = 15$

19. Solve for x. $\frac{x}{9} + 3 = 11$

20. $0.0009 \times 0.008 = ?$

1.	2.	3.	4.
5.	6.	7.	8.
9.	10.	11.	12.
13.	14.	15.	16.
17.	18.	19.	20.

Lesson #24

1. What value of x will make the equation true? $3x + 5 = 4x + 6$

2. Find the GCF of $10x^5y^4z$ and $30x^2y^2z$.

3. Find the circumference of this circle.

    8 cm

4. Solve for x. $\frac{x}{3} + 10 = -25$

5. How many inches are in 9 yards?

6. What is the value of x? $5(2x + 4) = 40$

7. Find the value of x. $6x - 4 = 14$

8. Simplify. $4(5a - 3b + 5) - 2(4a + 3)$

9. Find $\frac{4}{5}$ of 45.

10. $-54 + (-27) = ?$

11. $2.378 \times 0.3 = ?$

12. Simplify. $\dfrac{10x^2y^2z}{25xy^2}$

13. $35 \div 7 + 2[3(2) - 1] = ?$

14. $18\frac{2}{9} - 11\frac{7}{9} = ?$

15. Find the area of this trapezoid.

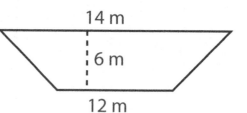

14 m

6 m

12 m

16. Solve for x. $7x - 5x + 9 = 21$

17. If $x = 2$ and $y = 3$, what is the value of $\frac{xy^2}{3} + 3$?

18. 30% of what number is 18?

19. $-|-79| = ?$

20. Give the coordinates of points A, B, and C.

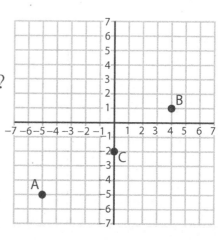

1.	2.	3.	4.
5.	6.	7.	8.
9.	10.	11.	12.
13.	14.	15.	16.
17.	18.	19.	20.

Lesson #25

1. $-96 - (-77) = ?$

2.   Find the LCM of $12ab^2c^5$ and $16a^2bc^2$.

3.   Solve for x. $\frac{x}{5} - 9 = 10$

4.   Full of sand, the wagon weighs 100 pounds. If the wagon weighs 80 ounces when it's empty, how much do the bags of sand weigh?

5.   Simplify. $\dfrac{3x^3y^2z}{15x^2yz^2}$

6.   Ten divided by the product of two times a number. Translate the phrase into an algebraic expression.

7.   Find the value of x. $5x - 55 = 25$

8. $3\frac{3}{5} + 6\frac{1}{4} = ?$

9.   Which is greater, $\frac{4}{5}$ or 95%?

10.   A ten-sided polygon is a(n) _____.

11.   Solve for x. $\frac{1}{8}x = 15$

12. $\frac{5}{6} \times \frac{9}{10} = ?$

13.   What is m when both sides of the equation are equal? $3m - 8 = 5m + 8$

14. $\frac{-840}{4} = ?$

15.   Write 48% as a decimal and as a reduced fraction.

16. $7 + 4[4 + 2(8 - 3) + 1] = ?$

17. $0.6 \div 0.12 = ?$

18.   Solve for x. $3(5x + 2) = -39$

19.   How many yards are in 3 miles?

20.   Find the measure of each angle.

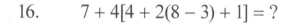

    $\angle$AOC _____ $\angle$AOE _____ $\angle$AOH _____

1.	2.	3.	4.
5.	6.	7.	8.
9.	10.	11.	12.
13.	14.	15.	16.
17.	18.	19.	20.

Lesson #26

1. A parking garage charges $5 for the first hour plus $1.50 for each additional half hour. What is the total cost for parking in the garage for 3 hours and 30 minutes?

2. Simplify. $7(4x - 3y + 9) - 4(2x + 3y - 4)$

3. Find the value of x. $-8x = 27 + x$

4. $34 + (-13) = ?$

5. Solve for x. $\frac{1}{5}x + 7 = 14$

6. How many quarts are 3 gallons?

7. What is the value of b? $b + 41 = 80$

8. $\frac{8}{9} \div \frac{1}{3} = ?$

9. Find the GCF of $8x^2yz^3$ and $12xyz^2$.

10. Solve the proportion to find x. $\frac{4}{9} = \frac{x}{117}$

11. Evaluate $3x - y + 9$ when $x = 3$ and $y = 2$.

12. $63 - 42\frac{4}{5} = ?$

13. Find the area of a circle if its radius is 5 cm long.

14. Draw perpendicular lines.

15. Simplify. $\frac{12a^3b^4c^2}{18a^2bc}$

16. Find the missing measure of angle x.

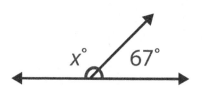

17. $0.005 \times 0.04 = ?$

18. Find the perimeter of a regular nonagon with sides of 8 inches each.

19. Solve for x. $3(3x + 2) = -21$

20. Give the coordinates of points D and F.

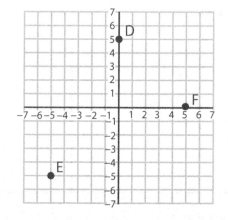

1.	2.	3.	4.
5.	6.	7.	8.
9.	10.	11.	12.
13.	14.	15.	16.
17.	18.	19.	20.

Lesson #27

1. Solve for y. $7y - 7 = 5y + 13$

2. Eight times a number is sixteen. Write it using algebraic symbols.

3. Find the P(H, T, H, H, H) on 5 flips of a coin.

4. $\begin{pmatrix} 4 & -5 \\ 0 & 7 \end{pmatrix} + \begin{pmatrix} -3 & 2 \\ 5 & -6 \end{pmatrix} = ?$

5. Find the area of a triangle with a base of 18 mm and a height of 4 mm.

6. $-3(-5)(2) = ?$

7. $5\frac{4}{9} + 6\frac{1}{2} = ?$

8. Write $\frac{7}{50}$ as a decimal.

9. How many yards are 108 inches?

10. Simplify. $\frac{4x^4 y^2}{8x^2 y}$

11. Solve for x. $\frac{x}{5} = 13$

12. Write the formula for finding the area of a parallelogram.

13. Find the LCM of $9a^3 b^2 c$ and $12a^2 bc$.

14. When $x = 8$, $y = 2$, and $z = 3$, what is the value of $\frac{x}{y} - z$?

15. Find the value of x. $x - 33 = -58$

16. $80,000 - 38,275 = ?$

17. There were 7 cantaloupes for every 8 watermelons in the garden. If there were 98 cantaloupes, how many watermelons were in the garden?

18. $2.5 \times 1.3 = ?$

19. Solve for x. $6(4x - 2) = 12$

20. Two angles whose measures add up to 180° are _____ angles.

1.	2.	3.	4.
5.	6.	7.	8.
9.	10.	11.	12.
13.	14.	15.	16.
17.	18.	19.	20.

Lesson #28

1.   Simplify. $6(3a - 2b + 4) + 2(5a - 5)$

2.   Draw a parallelogram.

3.   What value of x makes the equation true? $4x - 7 = 2x + 7$

4.   Find the value of 4^3.

5.   What is the surface area of a rectangular prism with a length of 9 feet, a width of 4 feet, and a height of 3 feet?

6.   What is the probability of rolling an odd number on one roll of a die?

7.   What is the value of $a(b + c)$ when $a = 4$, $b = 3$, and $c = 5$?

8.   Solve for x. $3x - 7 = 11$

9. $-74 - (-22) = ?$

10.   Solve for x. $\frac{x}{6} = 11$

11. $9 - 3\frac{2}{7} = ?$

12.   Find the GCF of $15x^2y^2z^4$ and $25xy^2z^3$.

13.   Find $\frac{3}{5}$ of 40.

14. $\frac{3}{7} \bigcirc \frac{5}{9}$

15.   What is the value of a? $a + 16 = -74$

16. $\frac{-123}{-3} = ?$

17. $7{,}566{,}244 + 9{,}895{,}355 = ?$

18.   Simplify. $\frac{10a^3b^2}{30a^2b}$

19.   Write $\frac{3}{5}$ as a decimal and as a percent.

20.   Give the point with each set of coordinates.

        a) $(5, -6)$ b) $(0, 1)$

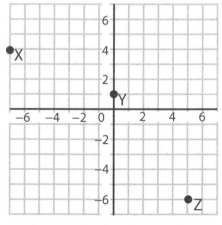

1.	2.	3.	4.
5.	6.	7.	8.
9.	10.	11.	12.
13.	14.	15.	16.
17.	18.	19.	20.

Lesson #29

1. How many feet are in 2 miles?

2. Which is greater, 0.65 or $\frac{1}{2}$?

3. $132 + (-68) = ?$

4. $1\frac{1}{5} \times \frac{5}{6} = ?$

5. Solve for a. $a + 17 = 62$

6. $7.230 \div 0.03 = ?$

7. Find the value of x. $\frac{x}{4} = 12$

8. Three times a number divided by two, decreased by four. Translate this sentence into an algebraic expression.

9. $\begin{pmatrix} -8 & 4 & 2 \\ -3 & 1 & -7 \end{pmatrix} + \begin{pmatrix} -1 & 3 & -5 \\ 0 & 9 & -6 \end{pmatrix} = ?$

10. Simplify. $\frac{4x^3y^2z}{12xy^2}$

11. Solve for x. $4x + 10 = 2x - 22$

12. Write $\frac{6}{50}$ as a decimal.

13. Round 6.396 to the nearest tenth.

14. Draw 2 parallel, horizontal lines.

15. Find the LCM of $6x^2y^3$ and $15xy^2z$.

16. On the Fahrenheit scale, water freezes at _____.

17. Solve for x. $7x - 8 = 13$

18. When $a = 5$ and $b = 3$, what is the value of $\frac{3a}{b} + 7$?

19. $30 - 4 \cdot 6 + 10 \div 2 = ?$

20. In the diagram, what is the missing measurement?

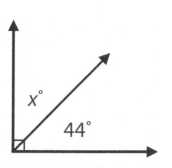

1.	2.	3.	4.
5.	6.	7.	8.
9.	10.	11.	12.
13.	14.	15.	16.
17.	18.	19.	20.

Lesson #30

1. $\frac{7}{8} \times \frac{12}{14} = ?$

2. Solve for a. $-6a = 72$

3. Write an algebraic expression that means a number divided by six.

4. What is the value of x? $\frac{x}{7} = -11$

5. What is the probability of rolling a number greater than 4 on one roll of a die?

6. Simplify. $\dfrac{12x^4y^3z^3}{18x^2yz^2}$

7. Solve for x. $4x - 6 = 6x + 14$

8. What is the area of the trapezoid?

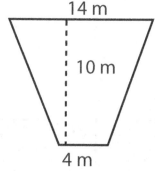

9. $-59 + (-37) = ?$

10. How many degrees are in a straight angle?

11. Simplify. $4(3a - 5b - 5) - 3(2a + 3b - 4)$

12. Find the LCM of $8x^2y^3$ and $14xy^2z^2$.

13. List the first four prime numbers.

14. Solve the proportion for x. $\frac{6}{8} = \frac{54}{x}$

15. Round 247,436,016 to the nearest ten million.

16. 90% of 40 is what number?

17. Write 3.479 using words.

18. What is the value of x? $4x - 8 = 12$

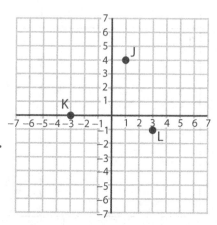

19. Write 0.42 as a percent and as a reduced fraction.

20. Give the coordinates of each point, J, and K.

1.	2.	3.	4.
5.	6.	7.	8.
9.	10.	11.	12.
13.	14.	15.	16.
17.	18.	19.	20.

Pre-Algebra

Mathematics
3rd Edition

Help Pages

Help Pages

Vocabulary

General
absolute value — the distance between a number, *x*, and zero on a number line; written as |*x*|. Example: |5| = 5 reads, "The absolute value of 5 is 5." |-7| = 7 reads, "The absolute value of -7 is 7."
composite number — a number with more than 2 factors. Example: 10 has factors of 1, 2, 5, and 10. Ten is a composite number.
exponent — tells the number of times that a base is multiplied by itself. An exponent is written to the upper right of the base. Example: 5^3 = 5 x 5 x 5. The exponent is 3.
expression — a mathematical phrase written in symbols. Example: $2x + 5$ is an expression.
factors — are multiplied together to get a product. Example: 2 and 3 are factors of 6.
greatest common factor (GCF) — the highest factor that 2 numbers have in common. Example: The factors of 6 are 1, 2, **3**, and 6. The factors of 9 are 1, **3**, and 9. The GCF of 6 and 9 is 3.
integers — the set of whole numbers, positive or negative, and zero.
least common multiple (LCM) — the smallest multiple that 2 numbers have in common. Example: Multiples of 3 are 3, 6, 9, **12**, 15... Multiples of 4 are 4, 8, **12**, 16... The LCM of 3 and 4 is 12.
multiples — can be evenly divided by a number. Example: 5, 10, 15, and 20 are multiples of 5.
prime factorization — a number, written as a product of its prime factors. Example: 140 can be written as 2 x 2 x 5 x 7 or 2^2 x 5 x 7.
prime number — a number with exactly 2 factors (the number itself and 1). 1 is not prime (it has only 1 factor). Example: 7 has factors of 1 and 7. Seven is a prime number.
square root — a number that when multiplied by itself gives you another number. The symbol for square root is $\sqrt{x}$. Example: $\sqrt{49} = 7$ reads, "The square root of 49 is 7."
term — the components of an expression, usually being added to or subtracted from each other. Example: The expression $2x + 5$ has two terms: $2x$ and 5. The expression $3n^2$ has only one term.
variable — a letter or symbol in an algebraic expression that represents a number.

Geometry
acute angle — an angle measuring less than 90°.
complementary angles — two angles whose measures add up to 90°.
congruent — figures with the same shape and the same size.
obtuse angle — an angle measuring more than 90°.
right angle — an angle measuring exactly 90°.
similar — figures having the same shape, but different sizes.
straight angle — an angle measuring exactly 180°.
supplementary angles — two angles whose measures add up to 180°. 
surface area — the sum of the areas of all of the faces of a solid figure.

Help Pages

Vocabulary (continued)

Geometry — Circles

circumference — the distance around the outside of a circle.

diameter — the widest distance across a circle. The diameter always passes through the center.

radius — the distance from any point on the circle to the center. The radius is half of the diameter.

Geometry — Polygons

Number of Sides		Name	Number of Sides		Name
3	△	triangle	7		heptagon
4	□	quadrilateral	8		octagon
5	⬠	pentagon	9		nonagon
6	⬡	hexagon	10		decagon

Geometry — Triangles

equilateral — a triangle in which all 3 sides have the same length.

isosceles — a triangle in which 2 sides have the same length.

scalene — a triangle in which no sides are the same length.

Measurement — Relationships

Volume	Distance
3 teaspoons in a tablespoon	36 inches in a yard
2 cups in a pint	1,760 yards in a mile
2 pints in a quart	5,280 feet in a mile
4 quarts in a gallon	100 centimeters in a meter
Weight	1,000 millimeters in a meter
16 ounces in a pound	**Temperature**
2,000 pounds in a ton	0°Celsius – freezing point
Time	100°Celsius – boiling point
10 years in a decade	32°Fahrenheit – freezing point
100 years in a century	212°Fahrenheit – boiling point

Ratio and Proportion

proportion — a statement that two ratios (or fractions) are equal. Example: $\frac{1}{2} = \frac{3}{6}$

percent (%) — the ratio of any number to 100. Example: 14% means 14 out of 100 or $\frac{14}{100}$.

Help Pages

Solved Examples

Absolute Value

The **absolute value** of a number is its distance from zero on a number line. It is always positive.

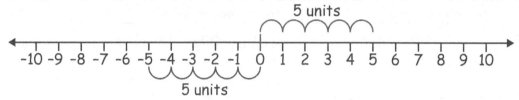

The absolute value of both -5 and +5 is 5, because both are 5 units away from zero. The symbol for the absolute value of -5 is |-5|. Examples: |-3| = 3; |8| = 8.

Coordinate Graphing

A **coordinate plane** is formed by the intersection of a horizontal number line, called the **x-axis**, and a vertical number line, called the **y-axis**. The axes meet at the point (0, 0), called the **origin**, and divide the coordinate plane into four **quadrants**.

Points are represented by **ordered pairs** of numbers, (x, y). The first number in an ordered pair is the x-coordinate; the second number is the y-coordinate. In the point (-4, 1), -4 is the x-coordinate, and 1 is the y-coordinate.

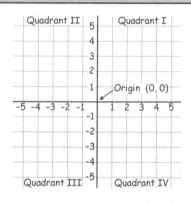

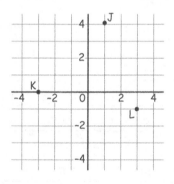

When graphing on a coordinate plane, always move on the x-axis first (right or left), and then move on the y-axis (up or down).

- The coordinates of point J are (1, 4).
- The coordinates of point K are (-3, 0).
- The coordinates of point L are (3, -1).

Decimals

Adding or subtracting decimals is very similar to adding or subtracting whole numbers. The main difference is that decimal points need to be lined-up before beginning.

Examples: Find the sum of 3.14 and 1.2.

```
  3.14
+ 1.20
------
  4.34
```

Add 55.1, 6.472, and 18.33.

```
 55.100
  6.472
+18.330
-------
 79.902
```

1. Line up the decimal points. Add zeroes as needed.
2. Add (or subtract) the decimals.
3. Add (or subtract) the whole numbers.
4. Bring the decimal point straight down.

Examples: Subtract 3.7 from 9.3.

```
  9.3
- 3.7
-----
  5.6
```

Find the difference of 4.1 and 2.88.

```
  4.10
- 2.88
------
  1.22
```

Help Pages

Solved Examples

Decimals (continued)

When **multiplying a decimal by a whole number**, the process is similar to multiplying whole numbers.

Examples: Multiply 3.42 by 4.   Find the product of 2.3 and 2.

3.42 → 2 decimal places
× 4 → 0 decimal places
13.68 → Place decimal point so there are 2 decimal places.

1. Line up the numbers on the right.
2. Multiply. Ignore the decimal point.
3. Place the decimal point in the product. The total number of decimal places in the product must equal the total number of decimal places in the factors.

2.3 → 1 decimal place
× 2 → 0 decimal places
4.6 → Place decimal point so there is 1 decimal place.

The process for **multiplying two decimal numbers** is a lot like the process described above.

Examples: Multiply 0.4 by 0.6.   Find the product of 2.67 and 0.3.

0.4 → 1 decimal place
× 0.6 → 1 decimal place
0.24 → Place decimal point so there are 2 decimal places.

2.67 → 2 decimal places
× 0.3 → 1 decimal place
0.801 → Place decimal point so there are 3 decimal places.

Sometimes it is necessary to **add zeroes in the product** as placeholders in order to have the correct number of decimal places.

Example: Multiply 0.03 by 0.4.

2 decimal places ⟶ 0.03
1 decimal place ⟶ × 0.4
Place decimal point so there are 3 decimal places. → 0.012

A zero had to be added in front of the 12, so there would be 3 decimal places in the product.

The process for **dividing a decimal number by a whole number** is similar to dividing whole numbers.

Examples: Divide 6.4 by 8.   Find the quotient of 20.7 and 3.

```
   0.8
8)6.4
 -6 4
    0
```

1. Set up the problem for long division.
2. Place the decimal point in the quotient directly above the decimal point in the dividend.
3. Divide. Add zeros as placeholders if necessary (examples below).

```
    6.9
3)20.7
 -18
   27
  -27
    0
```

Examples: Divide 4.5 by 6.   Find the quotient of 3.5 and 4.

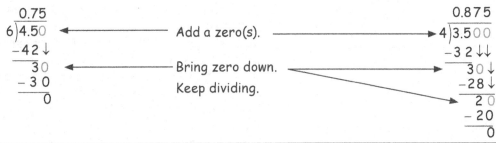

```
   0.75
6)4.50          ←  Add a zero(s).  →
 -42↓
   30           ←  Bring zero down.
  -30              Keep dividing.
    0
```

```
    0.875
4)3.500
 -32↓↓
   30↓
  -28↓
    20
   -20
     0
```

Help Pages

Solved Examples

Decimals (continued)

When dividing decimals, the remainder is not always zero. Sometimes, the division continues on and on, and the remainder begins to repeat itself. This quotient is called a **repeating decimal.**

Examples: Divide 2 by 3.   Divide 10 by 11.

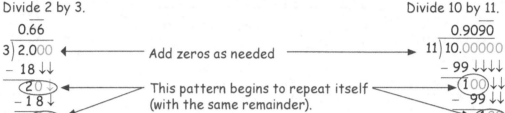

Add zeros as needed

This pattern begins to repeat itself (with the same remainder).

To write the final answer, put a bar in the quotient over the digits that repeat.

The process for **dividing a decimal number by a decimal number** is similar to other long division. The main difference is that the decimal point has to be moved in both the dividend and the divisor the <u>same number of places</u> to the right.

Example: Divide 1.8 by 0.3.   Divide 0.385 by 0.05.

$$
\begin{array}{r}
6. \\
0.3\overline{)1.8} \\
-18 \\
\hline
0
\end{array}
$$

1. Change the divisor to a whole number by moving the decimal point as many places to the right as possible.
2. Move the decimal in the dividend the same number of places to the right as the divisor.
3. Put the decimal point in the quotient directly above the decimal point in the dividend. Divide.

$$
\begin{array}{r}
7.7 \\
0.05\overline{)0.385} \\
-35 \\
\hline
35 \\
-35 \\
\hline
0
\end{array}
$$

Equations

An equation consists of two expressions separated by an equal sign: 2 + 3 = 5. More complicated equations involve variables, which replace a number. To solve an equation like this, figure out which number the variable stands for.
A simple example is when $2 + x = 5$. Here, the variable, x, stands for 3.

Sometimes an equation is not so simple. In these cases, there is a process for solving for the variable. No matter how complicated the equation, <u>the goal is to work with the equation until all the numbers are on one side and the variable is alone on the other side</u>. These equations will require only **one step** to solve. To check the answer, put the value of x back into the original equation.

Solving an **equation with a variable on one side:**

Example: Solve for x. $x + 13 = 27$

$$
\begin{array}{rcl}
x + 13 &=& 27 \\
-13 &=& -13 \\
\hline
x &=& 14
\end{array}
$$

Check: 14 + 13 = 27 ✓ correct!

Example: Solve for a. $a - 22 = -53$

$$
\begin{array}{rcl}
a - 22 &=& -53 \\
+22 &=& +22 \\
\hline
a &=& -31
\end{array}
$$

Check: -31 - 22 = -53 ✓ correct!

1. Look at the side of the equation that has the variable on it. If there is a number added to or subtracted from the variable, it must be removed. In the first example, 13 is added to x.
2. To remove 13, add its opposite (-13) to both sides of the equation.
3. Add downward. x plus nothing is x. 13 plus -13 is zero. 27 plus -13 is 14.
4. Once the variable is alone on one side of the equation, the equation is solved. The bottom line tells the value of x. $x = 14$.

Help Pages

Solved Examples

Equations (continued)

In the next examples, a number is either multiplied or divided by the variable (not added or subtracted).

Example: Solve for x. $3x = 39$

$$3x = 39$$

$$\frac{3x}{3} = \frac{39}{3}$$

$$x = 13$$

Check: $3(13) = 39$

$39 = 39$ ✓   correct!

Example: Solve for n. $\frac{n}{6} = -15$

$$\frac{n}{6}(6) = -15(6)$$

$$n = -90$$

Check: $\frac{-90}{6} = -15$

$-15 = -15$ ✓   correct!

1. Look at the side of the equation that has the variable on it. If there is a number multiplied by or divided into the variable, it must be removed. In the first example, 3 is multiplied by x.

2. To remove 3, divide both sides by 3. Use division because it is the opposite operation from the one in the equation (multiplication).

3. Follow the rules for multiplying or dividing integers. $3x$ divided by 3 is x. 39 divided by 3 is thirteen.

4. Once the variable is alone on one side of the equation, the equation is solved. The bottom line tells the value of x. $x = 13$.

The next set of examples also has a variable on only one side of the equation. These, however, are a bit more complicated because they require **two steps** in order to get the variable alone.

Example: Solve for x. $2x + 5 = 13$

$$2x + 5 = 13$$
$$\underline{-5 = -5}$$
$$2x = 8$$
$$\frac{2x}{2} = \frac{8}{2}$$
$$x = 4$$

Check: $2(4) + 5 = 13$

$8 + 5 = 13$

$13 = 13$ ✓   correct!

Example: Solve for n.

$$3n - 7 = 32$$
$$\underline{+7 = +7}$$
$$3n = 39$$
$$\frac{3n}{3} = \frac{39}{3}$$
$$n = 13$$

Check: $3(13) - 7 = 32$

$39 - 7 = 32$

$32 = 32$ ✓   correct!

1. Look at the side of the equation that has the variable on it. There is a number (2) multiplied by the variable, and there is a number added to it (5). Both of these must be removed. Always begin with the addition/subtraction. To remove the 5 add its opposite (-5) to both sides.

2. To remove the 2, divide both sides by 2. Divide, because it is the opposite operation from the one in the equation (multiplication).

3. Follow the rules for multiplying or dividing integers. $2x$ divided by 2 is x. 8 divided by 2 is four.

4. Once the variable is alone on one side of the equation, the equation is solved. The bottom line tells the value of x. $x = 4$.

Help Pages

Solved Examples

Equations (continued)

These multi-step equations also have a variable on only one side. To get the variable alone, though, requires several steps.

Example: Solve for x. $3(2x + 3) = 21$

$$\frac{\cancel{3}\left(\dfrac{2x+3}{\cancel{3}}\right)}{} = \frac{21}{3}$$

$$2x + 3 = 7$$
$$\underline{-3 = -3}$$
$$2x \quad = 4$$
$$\frac{\cancel{2}x}{\cancel{2}} = \frac{4}{2}$$
$$x = 2$$

Check: $3(2(2) + 3) = 21$

$$3(4 + 3) = 21$$
$$3(7) = 21$$
$$21 = 21 \checkmark \quad \text{correct!}$$

1. Look at the side of the equation that has the variable on it. First, the expression $(2x + 3)$ is multiplied by 3; then there is a number (3) added to $2x$, and there is a number (2) multiplied by x. All of these must be removed. To remove the 3 outside the parentheses, divide both sides by 3. Divide, because it is the opposite operation from the one in the equation (multiplication).
2. To remove the 3 inside the parentheses, add its opposite (-3) to both sides.
3. Remove the 2 by dividing both sides by 2.
4. Follow the rules for multiplying or dividing integers. $2x$ divided by 2 is x. 4 divided by 2 is two.
5. Once the variable is alone on one side of the equation, the equation is solved. The bottom line tells the value of x. $x = 2$.

When solving an **equation with a variable on both sides**, the goals are the same: to get the numbers on one side of the equation and to get the variable alone on the other side.

Example: Solve for x. $2x + 4 = 6x - 4$

$$2x + 4 = 6x - 4$$
$$\underline{-2x \quad\quad = -2x}$$
$$4 = 4x - 4$$
$$\underline{+4 = \quad\quad + 4}$$
$$8 = 4x$$
$$\frac{8}{4} = \frac{\cancel{4}x}{\cancel{4}}$$
$$2 = x$$

Check: $2(2) + 4 = 6(2) - 4$

$$4 + 4 = 12 - 4$$
$$8 = 8 \checkmark \quad \text{correct!}$$

1. Since there are variables on both sides, the first step is to remove the "variable term" from one of the sides by adding its opposite. To remove $2x$ from the left side, add $-2x$ to both sides.
2. There are also numbers added (or subtracted) to both sides. Next, remove the number added to the variable side by adding its opposite. To remove -4 from the right side, add +4 to both sides.
3. The variable still has a number multiplied by it. This number (4) must be removed by dividing both sides by 4.
4. The final line shows that the value of x is 2.

Help Pages

Solved Examples

Equations (continued)

Example: Solve for n. $5n - 3 = 8n + 9$

$$
\begin{array}{rcl}
5n - 3 &=& 8n + 9 \\
-8n & & = -8n \\
\hline
-3n - 3 &=& 9 \\
+3 &=& +3 \\
\hline
-3n &=& 12 \\
\dfrac{\cancel{3}n}{\cancel{3}} &=& \dfrac{12}{-3} \\
n &=& -4
\end{array}
$$

Check: $5(-4) - 3 = 8(-4) + 9$
$\quad\quad -20 - 3 = -32 + 9$
$\quad\quad\quad\quad -23 = -23 \checkmark$ correct!

Exponents

An **exponent** is a small number to the upper right of another number (the base). Exponents are used to show that the base is a repeated factor.

Example: 2^4 is read "two to the fourth power."
  The base (2) is a factor multiple times.
  The exponent (4) tells how many times the base is a factor.
  $2^4 = 2 \times 2 \times 2 \times 2 = 16$

$$\text{base} \longrightarrow 2^{4} \longleftarrow \text{exponent}$$

Example: 9^3 is read "nine to the third power" and means $9 \times 9 \times 9 = 729$.

Expressions

An **expression** is a number, a variable, or any combination of these, along with operation signs $(+, -, \times, \div)$ and grouping symbols. An expression never includes an equal sign.

Five examples of expressions are 5, x, $(x + 5)$, $(3x + 5)$, and $(3x^2 + 5)$.

To **evaluate an expression** means to calculate its value using specific variable values.

Example: Evaluate $2x + 3y + 5$ when $x = 2$ and $y = 3$.

$$2(2) + 3(3) + 5 = ?$$
$$4 + 9 + 5 = ?$$
$$13 + 5 = 18$$

The expression has a value of 18.

> 1. To evaluate, put the values of x and y into the expression.
> 2. Use the rules for integers to calculate the value of the expression.

Example: Find the value of $\dfrac{xy}{3} + 2$ when $x = 6$ and $y = 4$.

$$\frac{6(4)}{3} + 2 = ?$$
$$\frac{24}{3} + 2 = ?$$
$$8 + 2 = 10$$

The expression has a value of 10.

Help Pages

Solved Examples

Expressions (continued)

When evaluating a numerical expression containing multiple operations, use a set of rules called the **Order of Operations**. The Order of Operations determines which operations, and in which order, they should be performed. (Which operation should be done first, second, etc.)

The Order of Operations is as follows:
1. Parentheses
2. Exponents
3. Multiplication/Division (left to right in the order that they occur)
4. Addition/Subtraction (left to right in the order that they occur)

If parentheses are enclosed within other parentheses, work from the inside out.

To remember the order, use the mneumonic device "Please Excuse My Dear Aunt Sally."

Use the following examples to help you understand how to use the Order of Operations.

Example: $2 + 6 \cdot 5$

To evaluate this expression, work through the steps using the Order of Operations.
1. Since there are no parentheses or exponents in the expression, skip steps 1 and 2.
2. According to step 3, do multiplication and division. $6 \cdot 5 = 30$
3. Next, step 4 says to do addition and subtraction. $2 + 30 = 32$

The answer is 32.

Example: $42 \div 6 \cdot 3 + 4 - 16 \div 2$

$42 \div 6 \cdot 3 + 4 - 16 \div 2$
$7 \cdot 3 + 4 - 16 \div 2$
$21 + 4 - 16 \div 2$
$21 + 4 - 8$
$25 - 8$
17

1. Do multiplication and division first (in the order they occur).

2. Do addition and subtraction next (in the order they occur).

Example: $42 \div 6 \cdot 3 + 4 - 16 \div 2$

$5(2 + 4) + 15 \div (9 - 6)$
$5(6) + 15 \div (3)$
$30 + 5$
35

1. Do operations inside of parentheses first.

2. Do multiplication and division first (in the order they occur).

3. Do addition and subtraction next (in the order they occur).

Example: $4\left[3 + 2\left(7 + 5\right) - 7\right]$

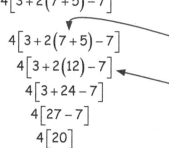

$4\left[3 + 2\left(7 + 5\right) - 7\right]$
$4\left[3 + 2(12) - 7\right]$
$4\left[3 + 24 - 7\right]$
$4\left[27 - 7\right]$
$4\left[20\right]$
80

1. Brackets are treated as parentheses. Start from the innermost parentheses first.

2. Then work inside the brackets.

Help Pages

Solved Examples

Expressions (continued)

Some expressions can be made more simple. There are a few processes for **simplifying an expression**. Deciding which process or processes to use depends on the expression itself. With practice, recognizing which of the following processes to use will become easier.

The **distributive property** is used when one term is multiplied by (or divided into) an expression that includes either addition or subtraction. $a(b+c) = ab + ac$ or $\dfrac{b+c}{a} = \dfrac{b}{a} + \dfrac{c}{a}$

Example: Simplify. $3(2x + 5)$

$$3(2x + 5) =$$
$$3(2x) + 3(5) =$$
$$6x + 15$$

1. Since the 3 is multiplied by the expression, $2x + 5$, the 3 must be multiplied by both terms in the expression.

2. Multiply 3 by $2x$, and then multiply 3 by $+ 5$.

Example: Simplify. $2(7x - 3y + 4)$

$$2(7x - 3y + 4) =$$
$$2(7x) + 2(-3y) + 2(+4) =$$
$$14x - 6y + 8$$

3. The result includes both of these: $6x + 15$. Notice that simplifying an expression does not result in a single number answer, only a more simple expression.

Expressions which contain like terms can also be simplified. **Like terms** are those that contain the same variable to the same power. $2x$ and $-4x$ are like terms; $3n^2$ and $8n^2$ are like terms; $5y$ and y are like terms; 3 and 7 are like terms.

An expression sometimes begins with like terms. This process for simplifying expressions is called **combining like terms**. When combining like terms, first identify the like terms. Then, simply add the like terms to each other, and write the results together to form a new expression.

Example: Simplify. $2x + 5y - 9 + 5x - 3y - 2$

> The like terms are $2x$ and $+5x$, $+5y$ and $-3y$, and -9 and -2.
>
> $2x + +5x = $**$+7x$**, $+5y + -3y = $**$+2y$**, and $-9 + -2 = $**$-11$**.
>
> The result is **$7x + 2y - 11$**.

The next examples are a bit more complex. It is necessary to use the distributive property first, and then to combine like terms.

Example: Simplify. $2(3x + 2y + 2) + 3(2x + 3y + 2)$

$$
\begin{array}{r}
6x + 4y + 4 \\
+6x + 9y + 6 \\
\hline
12x + 13y + 10
\end{array}
$$

1. First, apply the distributive property to each expression. Write the results on top of each other, lining up the like terms with each other. Pay attention to the signs of the terms.

Example: Simplify. $4(3x - 5y - 4) - 2(3x - 3y + 2)$

$$
\begin{array}{r}
+12x - 20y - 16 \\
-6x + 6y - 4 \\
\hline
6x - 14y - 20
\end{array}
$$

2. Then, add each group of like terms. Remember to follow the rules for integers.

Help Pages

Solved Examples

Expressions (continued)

Other expressions that can be simplified are written as fractions. **Simplifying** these expressions (**algebraic fractions**) is similar to simplifying numerical fractions. It involves cancelling out factors that are common to both the numerator and the denominator.

Simplify. $\dfrac{12x^2yz^4}{16xy^3z^2}$

$$\dfrac{\overset{3}{\cancel{12}}\ \cancel{x^2}^{x}\ \cancel{y}\ \cancel{z^4}^{z^2}}{\underset{4}{\cancel{16}}\ \cancel{x}\ y^3\ \cancel{z^2}_{y^2}}$$

$$\dfrac{\cancel{2}\cdot\cancel{2}\cdot3\cdot\cancel{x}\cdot x\cdot\cancel{y}\cdot\cancel{z}\cdot\cancel{z}\cdot z\cdot z}{\cancel{2}\cdot\cancel{2}\cdot2\cdot2\cdot\cancel{x}\cdot\cancel{y}\cdot y\cdot y\cdot\cancel{z}\cdot\cancel{z}}$$

$$\dfrac{3xz^2}{4y^2}$$

1. Begin by looking at the numerals in both the numerator and denominator (12 and 16). What is the largest number that goes into both evenly? Cancel this factor (4) out of both.

2. Look at the x portion of both numerator and denominator. What is the largest number of x's that can go into both of them? Cancel this factor (x) out of both.

3. Do the same process with y and then z. Cancel out the largest number of each (y and z^2). Write the numbers that remain in the numerator or denominator for the answer.

Often a relationship is described using verbal (English) phrases. In order to work with the relationship, first **translate it into an algebraic expression or equation**. In most cases, word clues will be helpful. Some examples of verbal phrases and their corresponding algebraic expressions or equations are written below.

Verbal Phrase	Algebraic Expression
Ten more than a number	$x + 10$
The sum of a number and five	$x + 5$
A number increased by seven	$x + 7$
Six less than a number	$x - 6$
A number decreased by nine	$x - 9$
The difference between a number and four	$x - 4$
The difference between four and a number	$4 - x$
Five times a number	$5x$
Eight times a number, increased by one	$8x + 1$
The product of a number and six is twelve.	$6x = 12$
The quotient of a number and 10	$\dfrac{x}{10}$
The quotient of a number and two, decreased by five	$\dfrac{x}{2} - 5$

In most problems, the word "is" means to put in an equal sign. When working with fractions and percents, the word "of" generally means multiply. Look at the example below.

One half <u>of</u> a number <u>is</u> fifteen.

Think of it as "one half <u>times</u> a number <u>equals</u> fifteen."

When written as an algebraic equation, it is $\dfrac{1}{2}x = 15$.

Help Pages

Solved Examples

Expressions (continued)

At times, finding the **greatest common factor (GCF) of an algebraic expression** is needed.

Example: Find the GCF of $12x^2yz^3$ and $18xy^3z^2$.

1. First, find the GCF of the numbers (12 and 18). The largest number that is a factor of both is **6**.
2. Now look at the x's. Of the x-terms, which contains fewer x's? Comparing x^2 and x, x has fewer.
3. Now look at the y's and then the z's. Again, of the y-terms, **y** contains fewer. Of the z-terms, z^2 has fewer.
4. The GCF contains all of these. **$6xyz^2$**

$\underline{12x^2yz^3 \text{ and } 18xy^3z^2}$

The GCF of 12 and 18 is **6**.

Of x^2 and x, the smaller is x.

Of y and y^3, the smaller is y.

Of z^3 and z^2, the smaller is z^2.

The GCF is $6xyz^2$.

At other times, finding the **least common multiple (LCM) of an algebraic expression** is called for.

Example: Find the LCM of $10a^3b^2c^2$ and $15ab^4c$.

1. First, find the LCM of the numbers (10 and 15). The lowest number that both go into evenly is **30**.
2. Look at the a-terms. Which has more a's? Comparing a^3 and a, a^3 has more.
3. Look at the b's and then the c's. Again, of the b-terms, b^4 contains more. Of the c-terms, c^2 contains more.
4. The LCM contains all of these. **$30a^3b^4c^2$**

$\underline{10a^3b^2c^2 \text{ and } 15ab^4c}$

The LCM of 10 and 15 is **30**. Of a^3 and a, the larger is a^3.

Of b^2 and b^4, the larger is b^4.

Of c^2 and c, the larger is c^2.

The LCM is $30a^3b^4c^2$.

Fractions

When **adding fractions that have different denominators**, first change the fractions so they have a common denominator. Then, add them.

Finding the **least common denominator (LCD)**: The LCD of the fractions is the same as the least common multiple of the denominators. Sometimes, the LCD will be the product of the denominators.

Example: Find the sum of $\frac{3}{8}$ and $\frac{1}{12}$.

```
2 | 8,12
2 | 4, 6
2 | 2,3
3 | 1,3
    1, 1
```

$2 \times 2 \times 2 \times 3 = 24$
The LCM is 24.

$\frac{3}{8} = \frac{9}{24}$

$+\frac{1}{12} = \frac{2}{24}$

$\frac{11}{24}$

1. First, find the LCM of 8 and 12.
2. The LCM of 8 and 12 is 24. This is also the LCD of these 2 fractions.
3. Find an equivalent fraction for each that has a denominator of 24.
4. When they have a common denominator, the fractions can be added.

Example: Add $\frac{1}{4}$ and $\frac{1}{5}$.

$4 \times 5 = 20$
The LCM is 20.

$\frac{1}{4} = \frac{5}{20}$

$+\frac{1}{5} = \frac{4}{20}$

$\frac{9}{20}$

Help Pages

Solved Examples

Fractions (continued)

When **adding mixed numbers with unlike denominators**, follow a process similar to the one used with fractions. Put the answer in simplest form.

Example: Find the sum of $6\frac{3}{7}$ and $5\frac{2}{3}$.

$$6\frac{3}{7} = 6\frac{9}{21}$$
$$+5\frac{2}{3} = 5\frac{14}{21} \qquad \frac{23}{21} = 1\frac{2}{21} + 11 = \boxed{12\frac{2}{21}}$$
$$11\frac{23}{21}$$
(improper)

1. Find the LCD.
2. Find the missing numerators.
3. Add the whole numbers, then add the fractions.
4. Make sure the answer is in simplest form.

When **subtracting numbers with unlike denominators**, follow a process similar to the one used when adding fractions. Put the answer in simplest form.

Examples: Find the difference of $\frac{3}{4}$ and $\frac{2}{5}$.

$$\frac{3}{4} = \frac{15}{20}$$
$$-\frac{2}{5} = \frac{8}{20}$$
$$\frac{7}{20}$$

1. Find the LCD just as when adding fractions.
2. Find the missing numerators.
3. Subtract the numerators and keep the common denominator.
4. Make sure the answer is in simplest form.

Subtract $\frac{1}{16}$ from $\frac{3}{8}$.

$$\frac{3}{8} = \frac{6}{16}$$
$$-\frac{1}{16} = \frac{1}{16}$$
$$\frac{5}{16}$$

When **subtracting mixed numbers with unlike denominators**, follow a process similar to the one used when adding mixed numbers. Put the answer in simplest form.

Example: Subtract $4\frac{2}{5}$ from $8\frac{9}{10}$.

1. Find the LCD.
2. Find the missing numerators.
3. Subtract and simplify the answer.

$$8\frac{9}{10} = 8\frac{9}{10}$$
$$-4\frac{2}{5} = 4\frac{4}{10}$$
$$4\frac{5}{10} = 4\frac{1}{2}$$

Sometimes when subtracting mixed numbers regrouping is necessary. If the numerator of the top fraction is smaller than the numerator of the bottom fraction, borrow from the whole number.

Example: Subtract $5\frac{5}{6}$ from $9\frac{1}{4}$.

1. Find the LCD.
2. Find the missing numerators. Write equivalent fractions.
3. Because 10 can't be subtracted from 3, rename the whole number as a mixed number using the common denominator.
4. Add the two fractions to get an improper fraction.
5. Subtract the whole numbers and the fractions, and simplify the answer.

$$9\frac{1}{4} = 9\frac{3}{12} = 8\frac{12}{12} + \frac{3}{12} = \boxed{8\frac{15}{12}}$$
$$-5\frac{5}{6} = 5\frac{10}{12} = \qquad 5\frac{10}{12}$$
$$3\frac{5}{12}$$

Help Pages

Solved Examples

Fractions (continued)

More examples of subtracting mixed numbers with unlike denominators:

$$8\frac{1}{2} = 8\frac{2}{4} = 7\frac{4}{4} + \frac{2}{4} = 7\frac{6}{4}$$
$$-4\frac{3}{4} = 4\frac{3}{4} = -4\frac{3}{4}$$
$$3\frac{3}{4}$$

$$10\frac{1}{5} = 10\frac{4}{20} = 9\frac{20}{20} + \frac{4}{20} = 9\frac{24}{20}$$
$$-6\frac{3}{4} = 6\frac{15}{20} = -6\frac{15}{20}$$
$$3\frac{9}{20}$$

To **multiply fractions**, multiply the numerators together to get the numerator of the product. Then, multiply the denominators together to get the denominator of the product. Make sure the answer is in simplest form.

Examples: Multiply $\frac{3}{5}$ by $\frac{2}{3}$.

$$\frac{3}{5} \times \frac{2}{3} = \frac{6}{15} = \frac{2}{5}$$

1. Multiply the numerators.
2. Multiply the denominators.
3. Simplify your answer.

Multiply $\frac{5}{8}$ by $\frac{4}{5}$.

$$\frac{5}{8} \times \frac{4}{5} = \frac{20}{40} = \frac{1}{2}$$

Sometimes cancelling works when multiplying fractions. Look at the examples again.

$$\frac{\cancel{3}^1}{5} \times \frac{2}{\cancel{3}_1} = \frac{2}{5}$$

1. Are there any numbers in the numerator and the denominator that have common factors?
2. If so, cross out the numbers, divide both by that factor, and write the quotient.
3. Then, multiply the fractions as described above, using the quotients instead of the original numbers.

$$\frac{\cancel{5}^1}{\cancel{8}_2} \times \frac{\cancel{4}^1}{\cancel{5}_1} = \frac{1}{2}$$

The 3s have a common factor — 3. Divide both of them by 3. Since, 3 ÷ 3 = 1, cross out the 3s, and write 1s in their place.

Now, multiply the fractions. In the numerator, 1 × 2 = 2. In the denominator, 5 × 1 = 5.

The answer is $\frac{2}{5}$.

As in the other example, the 5s can be cancelled.

But here, the 4 and the 8 also have a common factor (4).

8 ÷ 4 = 2 and 4 ÷ 4 = 1.

After cancelling both of these, multiply the fractions.

REMEMBER: It is possible to cancel up and down or diagonally, but NEVER sideways!

When **multiplying mixed numbers**, first change them into improper fractions.

Examples: Multiply $2\frac{1}{4}$ by $3\frac{1}{9}$.

$$2\frac{1}{4} \times 3\frac{1}{9} =$$
$$\frac{\cancel{9}^1}{\cancel{4}_1} \times \frac{\cancel{28}^7}{\cancel{9}_1} = \frac{7}{1} = 7$$

1. Change each mixed number to an improper fraction.
2. Cancel wherever possible.
3. Multiply the fractions.
4. Put the answer in simplest form.

Multiply $3\frac{1}{8}$ by 4.

$$3\frac{1}{8} \times 4 =$$
$$\frac{25}{\cancel{8}_2} \times \frac{\cancel{4}^1}{1} = \frac{25}{2} = 12\frac{1}{2}$$

Help Pages

Solved Examples

Fractions (continued)

To **divide fractions**, take the reciprocal of the 2nd fraction, and then multiply that reciprocal by the 1st fraction. Simplify the answer.

Examples: Divide $\frac{1}{2}$ by $\frac{7}{12}$.

$$\frac{1}{2} \div \frac{7}{12} =$$

$$\frac{1}{\cancel{2}_1} \times \frac{\cancel{12}^6}{7} = \frac{6}{7}$$

1. Keep the 1st fraction as it is.
2. Write the reciprocal of the 2nd fraction.
3. Change the sign to multiplication.
4. Cancel if possible and multiply.
5. Simplify the answer.

Divide $\frac{7}{8}$ by $\frac{3}{4}$.

$$\frac{7}{8} \div \frac{3}{4} =$$

$$\frac{7}{\cancel{8}_2} \times \frac{\cancel{4}^1}{3} = \frac{7}{6} = 1\frac{1}{6}$$

When **dividing mixed numbers**, first change them into improper fractions.

Example: Divide $1\frac{1}{4}$ by $3\frac{1}{2}$.

$$1\frac{1}{4} \div 3\frac{1}{2} =$$

$$\frac{5}{4} \div \frac{7}{2} =$$

$$\frac{5}{\cancel{4}_2} \times \frac{\cancel{2}^1}{7} = \frac{5}{14}$$

1. Change each mixed number to an improper fraction.
2. Keep the 1st fraction as it is.
3. Write the reciprocal of the 2nd fraction.
4. Change the sign to multiplication.
5. Cancel if possible and multiply.
6. Simplify the answer.

Geometry

To find the **area of a triangle**, first recognize that any triangle is exactly half of a parallelogram.

The whole figure is a parallelogram.

Half of the whole figure is a triangle.

The triangle's area is equal to half of the product of the base and the height.

$$\text{Area of triangle} = \frac{1}{2}(\text{base} \times \text{height}) \quad \text{or} \quad A = \frac{1}{2}bh$$

Examples: Find the area of the triangles below.

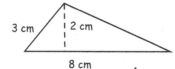

3 cm   2 cm   8 cm

So, $A = 8 \text{ cm} \times 2 \text{ cm} \times \frac{1}{2} = \textbf{8 cm}^2$

1. Find the length of the base. (8 cm)
2. Find the height. (It is 2 cm. The height is always straight up and down—never slanted.)
3. Multiply them together and divide by 2 to find the area. (8 cm²)

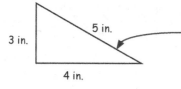

3 in.   5 in.   4 in.

The base of this triangle is 4 inches long. Its height is 3 inches. (Remember the height is always straight up and down!)

So, $A = 4 \text{ in.} \times 3 \text{ in.} \times \frac{1}{2} = \textbf{6 in.}^2$

Help Pages

Solved Examples

Geometry

Finding the **area of a parallelogram** is similar to finding the area of any other quadrilateral. The area of the figure is equal to the length of its base multiplied by the height of the figure.

Area of parallelogram = base × height   or $A = b \times h$

Example: Find the area of the parallelogram below.

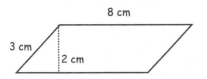

1. Find the length of the base. (8 cm)
2. Find the height. (It is 2 cm. The height is always straight up and down—never slanted.)
3. Multiply to find the area. (16 cm²)

So, $A = 8$ cm $\times 2$ cm = **16 cm²**

Finding the **area of a trapezoid** is a little different from other quadrilaterals. Trapezoids have 2 bases of unequal length. To find the area, first find the average of the lengths of the 2 bases. Then, multiply that average by the height.

Area of trapezoid = $\dfrac{base_1 + base_2}{2} \times height$ or $A = (\dfrac{b_1 + b_2}{2})h$

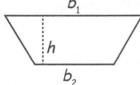

The bases are labeled b_1 and b_2.

The height, h, is the distance between the bases.

Example: Find the area of the trapezoid below.

1. Add the lengths of the two bases. (22 cm)
2. Divide the sum by 2. (11 cm)
3. Multiply that result by the height to find the area. (110 cm²)

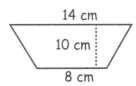

$\dfrac{14 \text{ cm} + 8 \text{ cm}}{2} = \dfrac{22 \text{ cm}}{2} = 11$ cm

11 cm $\times 10$ cm = **110 cm²** = Area

To **find the measure of an angle**, a protractor is used.

The symbol for angle is ∠. On the diagram, ∠AOE has a measure less than 90°, so it is acute.

With the center of the protractor on the vertex of the angle (where the 2 rays meet), place one ray ($\overrightarrow{OA}$) on one of the "0" lines. Look at the number that the other ray ($\overrightarrow{OE}$) passes through. Since the angle is acute, use the lower set of numbers. Since $\overrightarrow{OE}$ is halfway between the 40 and the 50, the measure of ∠AOE is 45°. (If it were an obtuse angle, the higher set of numbers would be used.)

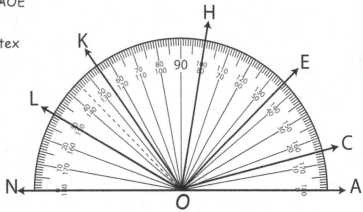

Look at ∠NOH. It is an obtuse angle, so the higher set of numbers will be used. Notice that $\overrightarrow{ON}$ is on the "0" line. $\overrightarrow{OH}$ passes through the 100 mark. The measure of ∠NOH is 100°.

Help Pages

Solved Examples

Geometry (continued)

The **circumference of a circle** is the distance around the outside of the circle. Before finding the circumference of a circle either its radius or its diameter must be known. The value of the constant, pi (π) $\pi = 3.14$ (rounded to the nearest hundredth) must be known.

With this information, the circumference can be found by multiplying the diameter by pi.

Circumference = $\pi \times$ diameter   or $C = \pi d$

Examples: Find the circumference of the circles below.

1. Find the length of the diameter. (12 m)
2. Multiply the diameter by π. (12 m × 3.14)
3. The product is the circumference. (37.68 m)

$C = 12$ m × 3.14 = **37.68 m**

Sometimes the radius of a circle is given instead of the diameter. The radius of any circle is exactly half of the diameter. If a circle has a radius of 3 feet, its diameter is 6 feet.

1. Since the radius is 4 mm, the diameter must be 8 mm.
2. Multiply the diameter by π. (8 mm × 3.14)
3. The product is the circumference. (25.12 mm)

$C = 8$ mm × 3.14 = **25.12 mm**

When finding the **area of a circle**, the length of the radius is squared (multiplied by itself), and then that answer is multiplied by the constant, pi (π). $\pi = 3.14$ (rounded to the nearest hundredth).

Area = $\pi \times$ radius × radius or $A = \pi r^2$

Examples: Find the area of the circles below.

1. Find the length of the radius. (9 mm)
2. Multiply the radius by itself. (9 mm x 9 mm)
3. Multiply the product by pi. (81 mm² x 3.14)
4. The result is the area. (254.34 mm²)

$A = 9$ mm x 9 mm x 3.14 = **254.34 mm²**

Sometimes the diameter of a circle is given instead of the radius. Remember, the diameter of any circle is exactly twice the radius. If a circle has a diameter of 6 feet, its radius is 3 feet.

1. Since the diameter is 14 m, the radius must be 7 m.
2. Square the radius. (7 m x 7 m)
3. Multiply that result by π. (49 m² × 3.14)
4. The product is the area. (153.86 m²)

$A = (7$ m$)^2$ x 3.14 = **153.86 m²**

Help Pages

Solved Examples

Geometry (continued)

To find the **surface area** of a solid figure, first count the total number of faces. Then, find the area of each of the faces; finally, add the areas of each face. That sum is the surface area of the figure.

Here, the focus will be on finding the **surface area of a rectangular prism**. A rectangular prism has 6 faces. Actually, the opposite faces are identical, so this figure has 3 pairs of faces. Also, a prism has only 3 dimensions: length, width, and height.

This prism has identical left and right sides (A and B), identical top and bottom (C and D), and identical front and back (unlabeled).

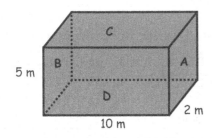

1. Find the area of the front: *l* x *w*. (10 m x 5 m = 50 m²) Since the back is identical, its area is the same.

2. Find the area of the top (C): *l* x *h*. (10 m x 2 m = 20 m²) Since the bottom (D) is identical, its area is the same.

3. Find the area of side A: *w* x *h*. (2 m x 5 m = 10 m²) Since side B is identical, its area is the same.

4. Add up the areas of all 6 faces.

   (10 m² + 10 m² + 20 m² + 20 m² + 50 m² + 50 m² = 160 m²)

The formula is Surface Area = 2(length x width) + 2(length x height) + 2(width x height)

or SA = 2*lw* + 2*lh* + 2*wh*

Ordering Integers

Integers include the counting numbers, their opposites (negative numbers), and zero.

The negative numbers are to the left of zero.   The positive numbers are to the right of zero.

When **ordering integers**, they are being arranged either from least to greatest or from greatest to least. The further a number is to the right, the greater its value. For example, 9 is further to the right than 2, so 9 is greater than 2.

In the same way, -1 is further to the right than -7, so -1 is greater than -7.

Examples: Order these integers from **least to greatest**.   -10, 9, -25, 36, 0

Remember, the smallest number will be the one farthest to the left on the number line, -25, then -10, then 0. Next will be 9, and finally 36.

Answer: -25, -10, 0, 9, 36

Put these integers in order from **greatest to least**. -94, -6, -24, -70, -14
Now the greatest value (farthest to the right) will come first, and the smallest value (farthest to the left) will come last.

Answer: -6, -14, -24, -70, -94

Help Pages

Solved Examples

Integers (continued)

The rules for performing operations $(+, -, \times, \div)$ on integers are very important and must be memorized.

The Addition Rules for Integers:

1. When the signs are the same, add the numbers and keep the sign.

$$
\begin{array}{r} +33 \\ + +19 \\ \hline +52 \end{array} \qquad\qquad \begin{array}{r} -33 \\ + -19 \\ \hline -52 \end{array}
$$

2. When the signs are different, subtract the numbers and use the sign of the larger number.

$$
\begin{array}{r} +33 \\ + -19 \\ \hline +14 \end{array} \qquad\qquad \begin{array}{r} -55 \\ + +27 \\ \hline -28 \end{array}
$$

The Subtraction Rule for Integers:

Change the sign of the second number and add (follow the Addition Rule for Integers above).

$$
\begin{array}{r} +56 \\ - -26 \\ \hline \end{array} \xrightarrow{\text{apply rule}} \begin{array}{r} +56 \\ + +26 \\ \hline +82 \end{array} \qquad \begin{array}{r} +48 \\ - +23 \\ \hline \end{array} \xrightarrow{\text{apply rule}} \begin{array}{r} +48 \\ + -23 \\ \hline +25 \end{array}
$$

Notice that every subtraction problem becomes an addition problem, using this rule!

The Multiplication and Division Rule for Integers:

1. When the signs are the same, the answer is positive (+).

$$+7 \times +3 = +21 \qquad\qquad -7 \times -3 = +21$$

$$+18 \div +6 = +3 \qquad\qquad -18 \div -6 = +3$$

2. When the signs are different, the answer is negative (-).

$$+7 \times -3 = -21 \qquad\qquad -7 \times +3 = -21$$

$$-18 \div +6 = -3 \qquad\qquad +18 \div -6 = -3$$

The chart to the right contains a helpful summary of this rule.

+		+		+
-	$\times$	-		+
+		-	=	-
-		+		-
+		+		+
-	$\div$	-		+
+		-		-
-		+		-

Proportion

A **proportion** is a statement that two ratios are equal to each other. There are two ways to solve a proportion when a number is missing.

1. One way to solve a proportion is already familiar. Use the equivalent fraction method.

$$\overset{\times 8}{\frown}$$
$$\frac{5}{8} = \frac{n}{64}$$
$$\underset{\times 8}{\smile}$$

$$n = 40$$

So, $\dfrac{5}{8} = \dfrac{40}{64}$.

To use Cross-Products:
1. Multiply downward on each diagonal.
2. Make the product of each diagonal equal to each other.
3. Solve for the missing variable.

2. Another way to solve a proportion is by using cross-products.

$$\frac{14}{20} = \frac{21}{n}$$

$$20 \times 21 = 14 \times n$$

$$420 = 14n$$

$$\frac{420}{14} = \frac{14n}{14}$$

$$30 = n$$

So, $\dfrac{14}{20} = \dfrac{21}{30}$.

Help Pages

Solved Examples

Percent

When changing from a fraction to a percent, a decimal to a percent, or from a percent to either a fraction or a decimal, it is very helpful to use an FDP chart (Fraction, Decimal, Percent).

To change a **fraction to a percent and/or decimal**, first find an equivalent fraction with 100 in the denominator. Once the equivalent fraction is found, it can easily be written as a decimal. To change that decimal to a percent, move the decimal point 2 places to the right and add a % sign.

Example: Change $\frac{2}{5}$ to a percent, and then to a decimal.

1. Find an equivalent fraction with 100 in the denominator.
2. From the equivalent fraction above, the decimal can easily be found. Say the name of the fraction: "forty hundredths." Write this as a decimal. 0.40
3. To change 0.40 to a percent, move the decimal two places to the right. Add a % sign. 40%

F	D	P
$\frac{2}{5}$		

F	D	P
$\frac{2}{5} = \frac{?}{100}$	0.40	

F	D	P
$\frac{2}{5} = \frac{?}{100}$	0.40	40%

$$\overset{\times 20}{\frac{2}{5}} = \underset{\times 20}{\frac{?}{100}}$$

? = 40

$$\frac{2}{5} = \frac{40}{100} = 0.40$$

$0.40 = 40\%$

When changing from a **percent to a decimal or a fraction**, the process is similar to the one used on the page above. Begin with the percent. Write it as a fraction with a denominator of 100; reduce this fraction. Return to the percent, move the decimal point 2 places to the left. This is the decimal.

Example: Write 45% as a fraction, and then as a decimal.

1. Begin with the percent. (45%) Write a fraction the denominator is 100, and the numerator is the "percent." $\frac{45}{100}$

2. This fraction must be reduced. The reduced fraction is $\frac{9}{20}$.

3. Go back to the percent. Move the decimal point two places to the left to change it to a decimal.

$$45\% = \frac{45}{100}$$

$$\frac{45(\div 5)}{100(\div 5)} = \frac{9}{20}$$

$$45\% = .45$$

Decimal point goes here.

When changing from a **decimal to a percent or a fraction**, again, the process is similar to the one used above. Begin with the decimal. Move the decimal point 2 places to the right, and add a % sign. Return to the decimal. Write it as a fraction and reduce.

Example: Write 0.12 as a percent, and then as a fraction.

1. Begin with the decimal. (0.12) Move the decimal point two places to the right to change it to a percent.

2. Go back to the decimal, and write it as a fraction. Reduce this fraction.

$0.12 = 12\%$

0.12 = twelve hundredths

$$\frac{12}{100} = \frac{12(\div 4)}{100(\div 4)} = \frac{3}{25}$$

Help Pages

Solved Examples

Compound Probability

The **probability of two or more independent events** occurring together can be determined by multiplying the individual probabilities together. The product is called the compound probability.

> Probability of A & B = (Probability of A) x (Probability of B)
>
> or   P(A and B) = P(A) x P(B)

Example: What is the probability of rolling a 6 and then a 2 on two rolls of a die [P(6 and 2)]?

A) First, find the probability of rolling a 6 [P(6)]. Since there are 6 numbers on a die and only one of them is a 6, the probability of getting a 6 is $\frac{1}{6}$.

B) Then find the probability of rolling a 2 [P(2)]. Since there are 6 numbers on a die and only one of them is a 2, the probability of getting a 2 is $\frac{1}{6}$.

So, P(6 and 2) = P(6) x P(2) = $\frac{1}{6} \times \frac{1}{6} = \frac{1}{36}$

There is a 1 to 36 chance of getting a 6 and then a 2 on two rolls of a die.

Example: What is the probability of getting a 4 and then a number greater than 2 on two spins of this spinner [P(4 and greater than 2)]?

A) First, find the probability of getting a 4 [P(4)]. Since there are 4 numbers on the spinner and only one of them is a 4, the probability of getting a 4 is $\frac{1}{4}$.

B) Then find the probability of getting a number greater than 2 [P(greater than 2)]. Since there are 4 numbers on the spinner and two of them are greater than 2, the probability of getting a 2 is $\frac{2}{4}$.

So, P(2 and greater than 2) = P(2) x P(greater than 2) = $\frac{1}{4} \times \frac{2}{4} = \frac{2}{16} = \frac{1}{8}$.

There is a 1 to 8 chance of getting a 4 and then a number greater than 2 on two spins of a spinner.

Example: On three flips of a coin, what is the probability of getting heads, tails, heads [P(H,T,H)]?

A) First, find the probability of getting heads [P(H)]. Since there are only 2 sides on a coin and only one of them is heads, the probability of getting heads is $\frac{1}{2}$.

B) Then find the probability of getting tails [P(T)]. Again, there are only 2 sides on a coin and only one of them is tails. The probability of getting tails is also $\frac{1}{2}$.

So, P(H,T,H) = P(H) x P(T) x P(H) = $\frac{1}{2} \times \frac{1}{2} \times \frac{1}{2} = \frac{1}{8}$

There is a 1 to 8 chance of getting heads, tails and then heads on 3 flips of a coin.

Who Knows?

Degrees in a right angle? ....(90)

A straight angle? ....(180)

Angle greater than 90°? ....(obtuse)

Less than 90°? ....(acute)

Sides in a quadrilateral? ....(4)

Sides in an octagon?...........................(8)

Sides in a hexagon? ....(6)

Sides in a pentagon? ....(5)

Sides in a heptagon? ....(7)

Sides in a nonagon? ....(9)

Sides in a decagon?.......................... (10)

Inches in a yard? ....(36)

Yards in a mile? ....(1,760)

Feet in a mile? ....(5,280)

Centimeters in a meter? ....(100)

Teaspoons in a tablespoon? .... (3)

Ounces in a pound? ....(16)

Pounds in a ton?...........................(2,000)

Cups in a pint? ....(2)

Pints in a quart? ....(2)

Quarts in a gallon? ....(4)

Millimeters in a meter? .... (1,000)

Years in a century? ....(100)

Years in a decade? ....(10)

Celsius freezing? ....(0°C)

Celsius boiling? ....(100°C)

Fahrenheit freezing? ....(32°F)

Fahrenheit boiling? ....(212°F)

Number with only 2 factors? .... (prime)

Perimeter? ....(add the sides)

Area? ....(length x width)

Volume? .... (length x width x height)

Area of parallelogram?...........................
.... (base x height)

Area of triangle? ....($\frac{1}{2}$ base x height)

Area of trapezoid?...................................
..............................($\frac{base + base}{2} \times height$)

Surface area of a rectangular
prism? ....2(lw) + 2(wh) + 2(lh)

Area of a circle? ....(πr^2)

Circumference of a circle? ....(πd)

Triangle with no sides equal? ....
.... (scalene)

Triangle with 3 sides equal?..................
....(equilateral)

Triangle with 2 sides equal? ....
.... (isosceles)

Distance across the middle of a circle?
.... (diameter)

Half of the diameter? .... (radius)

Figures with the same size
and shape? ....(congruent)

Figures with same shape,
different sizes? ....(similar)

Number occurring most often? ....
.... (mode)

Middle number? ....(median)

Answer in addition? ....(sum)

Answer in division? ....(quotient)

Answer in multiplication? ....(product)

Answer in subtraction? ......(difference)

Pre-Algebra

Mathematics
3rd Edition

Answers to Lessons

	Lesson #1		Lesson #2		Lesson #3
1	$\frac{1}{2}$	1	$x = 14$	1	$a = -20$
2	$\frac{1}{216}$	2	72 mm²	2	476 miles
3	12.357	3	35	3	$5\frac{3}{5}$
4	$20a - 8b + 10$	4	52	4	$x = 81$
5	$2\frac{1}{6}$	5	$26\frac{11}{15}$	5	0.85, $\frac{17}{20}$
6	18	6	$x = 98$	6	84 cm²
7	-49	7	141°	7	127
8	-72	8	$a = -61$	8	$10x + 21y - 32$
9	120	9	0.14, 14%	9	congruent
10	>	10	$4 = x$	10	198 in.²
11	64 quarts	11	6	11	3
12	$x = 91$	12	$8 - 6x$	12	1.02
13	$a = -4$	13	21,120 feet	13	$5a^3b^2c$
14	$x = -29$	14	$a = 64$	14	72
15	0.18, $\frac{9}{50}$	15	-36	15	$\frac{1}{6}$
16	4,382	16	⬡　　⬡	16	$x = -11$
17	180 in.²	17	47.002	17	$2x + 4 = 12$
18	$x = -5$	18	cup, pint, quart, gallon	18	80%
19	$x = 36$	19	$\frac{5}{8}$, 5 to 8	19	$x = -9$
20	$\frac{5}{9}$	20	$2,700	20	complementary

	Lesson #4		Lesson #5		Lesson #6
1	−29	1	$y = 5$	1	46.3
2	0.156	2	16	2	$x = -6$
3	900 centimeters	3	18	3	$a = -12$
4	scalene	4	$7x - 10$	4	80%
5	$18\frac{2}{9}$	5	$a = 6$	5	−48
6	1,201	6	35	6	3 gallons
7	6	7	$P = 36\text{cm}, A = 81\text{cm}^2$	7	$1\frac{1}{2}$
8	$a = 83$	8	16	8	$a = 78$
9	22	9	$3 \cdot 3 \cdot 5 \cdot x \cdot x \cdot y \cdot y$	9	0.729
10	$3\frac{1}{5}$	10	$\frac{4}{9}$	10	
11	$x = -6$	11	1	11	<
12	45	12	27,183	12	$\frac{x}{6}$
13	$5 = a$	13	1:30 pm	13	25
14	64 inches	14	$9\frac{11}{12}$	14	32
15	$x = 108$	15	−61, −44, −16, −2	15	$a = -77$
16	37.68 cm	16	$x = 2$	16	$48a^3b^4c^2$
17	2	17	$x = 48$	17	0.15, 15%
18	60,000,000	18	thirteen and one thousandth	18	27,408
19	0.000035	19	212°F	19	420 ft²
20	16 cups	20	53°	20	10 sides

	Lesson #7		Lesson #8		Lesson #9
1	32	**1**	$6xy^2z^3$	**1**	16
2	1,000,000,000	**2**	37	**2**	$x = 1$
3	-63	**3**	-192	**3**	-42
4	$17\frac{5}{6}$	**4**	14 pints	**4**	24 cm^2
5	78 bags of chips	**5**	$a = -14$	**5**	$b = 108$
6	$x = 3$	**6**	$3x$	**6**	96 ounces
7	101°	**7**	⟷ ⟷	**7**	130
8	230.27	**8**	$a = 38$	**8**	$x = 143$
9	$27a + 18b + 27c - 45$	**9**	3	**9**	$18x + 12y - 21z - 12$
10	22	**10**	$11\frac{4}{7}$	**10**	$x = -4$
11	0.125	**11**	$x = -140$	**11**	$\frac{1}{36}$
12	$x = 110$	**12**	supplementary	**12**	100°C
13	$a = -4$	**13**	36	**13**	625
14	1,760 yards	**14**	44	**14**	33
15	$x = 189$	**15**	80 ft^3	**15**	$2 \cdot 2 \cdot 5 \cdot a \cdot a \cdot b$
16	9.6 cm	**16**	0.000064	**16**	$b = -15$
17	2.2, 2.12, 2.02, 2.001	**17**	$\frac{1}{3}$	**17**	0.65; $\frac{13}{20}$
18	153.86 cm^2	**18**	$x = 7$	**18**	similar
19	160	**19**	3:4; 3 to 4	**19**	10.57
20	149	**20**	$x = 35$	**20**	$\angle AOD = 25°$ $\angle GOE = 55°$

	Lesson #10		Lesson #11		Lesson #12
1	113	1	$x = 72$	1	43.96 in.
2	$x + 12$	2	$4\frac{3}{7}$	2	0.2544
3	$x = 4$	3	27 feet	3	6 tons
4	8,000 pounds	4	parallelogram	4	$7 = x$
5	216 m²	5	$a = -207$	5	$12\frac{5}{8}$
6	3,147	6	$<$	6	$a = -134$
7	49	7	$\frac{2}{3}$	7	90 Quarter horses
8	25	8	$x = 13$	8	-121
9	$120a^4b^2$	9	0.083	9	$x = 7$
10	$x = -135$	10	375 oranges	10	$6a^2b$
11	-24	11	102	11	11
12	$b = 12$	12	product	12	0.06; 6%
13	500 years	13	6.8	13	$x = 88$
14	-59	14	$\frac{7}{x} - 3$	14	74°
15	15	15	37,332	15	$<$
16	difference	16	180°	16	30%
17	$\frac{3}{5}$	17	$1\frac{1}{4}$	17	190
18	$\frac{8}{25}$; 32%	18	$a = -167$	18	
19	$x = -10$	19	$6 = a$	19	isosceles
20	$1.80	20	equilateral	20	$x = -21$

	Lesson #13		Lesson #14		Lesson #15
1	$6a - 20b + 2$	**1**	302 pounds	**1**	mode
2	6	**2**	0.00072	**2**	$13\frac{7}{15}$
3	12	**3**	$a = -124$	**3**	$a = 80$
4	105	**4**	60 quarts	**4**	$x = -100$
5	20	**5**	54	**5**	144 mm²
6	$\frac{2}{9}$	**6**	0	**6**	32
7	60	**7**	$x = 112$	**7**	$56x^5y^3z^2$
8	12 cups	**8**	136 m²	**8**	112 ounces
9	75%	**9**	18	**9**	$a = 3$
10	335,029	**10**	$x = -3$	**10**	43.26
11	$a = -72$	**11**	$a = -26$	**11**	51
12	$\frac{12x}{2}$	**12**	$0.44; \frac{11}{25}$	**12**	⬡ ⬡
13	95.63	**13**	$8\frac{1}{4}$	**13**	44
14	$x = 51$	**14**	-32	**14**	$x = 15$
15	2,666	**15**	14	**15**	$\frac{x}{5} + 9$
16	$<$	**16**	$b = 104$	**16**	$7 = a$
17	$x = 3$	**17**	750,000,000	**17**	4^5
18	$x = -96$	**18**	$A = \frac{1}{2}bh$	**18**	$\frac{1}{32}$
19	2, 3, 5, 7, 11	**19**	$2 \cdot 3 \cdot 5 \cdot a \cdot a \cdot b \cdot b \cdot b \cdot c$	**19**	1,772,015,040
20	$x = 5$	**20**	$x = 350$	**20**	$\frac{11}{20}$

	Lesson #16		**Lesson #17**		**Lesson #18**
1	$1 = b$	**1**	0.28	**1**	$x = 14$
2	-87	**2**	$11\frac{7}{8}$	**2**	$c = 194$
3	78	**3**	$a = 25$	**3**	$5xyz$
4	$3\frac{2}{5}$	**4**	48	**4**	1,183
5	$2 \cdot 3 \cdot 3 \cdot x \cdot x \cdot y$	**5**	61	**5**	$a = 7$
6	$\frac{2}{3}$	**6**	0.000035	**6**	$-25, -12, 0, 18$
7	5 centuries	**7**	$<$	**7**	24
8	$a = -34$	**8**	28%	**8**	$60a^4b^2$
9	17	**9**	$11 - x$	**9**	$x = 42$
10	40	**10**	scalene	**10**	$6x - 8$
11	74.942	**11**	1	**11**	16
12	7 pounds 8 ounces	**12**	19	**12**	42.32
13	$a = -19$	**13**	$8x - 14y + 10$	**13**	$<$
14	4,815,406	**14**	28.26 m²	**14**	-70
15	-220	**15**	$1\frac{1}{2}$	**15**	30%
16	$b = -41$	**16**	$2 \cdot 2 \cdot 5 \cdot x \cdot x \cdot y \cdot y \cdot y \cdot z$	**16**	286 cm²
17	7	**17**	$x = 6$	**17**	$x = 7$
18	38%; $\frac{19}{50}$	**18**	$x = -40$	**18**	-34
19	11 ft	**19**	25	**19**	$x = 9.5$
20	3 inches	**20**	$x = 52$	**20**	F(−3,2); G(1, −2); H(2, 0)

	Lesson #19		Lesson #20		Lesson #21
1	$\frac{2}{3}$	1	$3ab^2c$	1	$C = d\pi$ or $2\pi r$
2	26	2	$x = -2$	2	$14a + 27b - 19$
3	180°	3	$18x + 2y + 1$	3	
4	27	4	$y = -95$	4	18
5	$\frac{1}{216}$	5	0.252	5	$a = -6$
6	0.076	6	$x = -28$	6	39
7	0.16, 16%	7	$2 \cdot 3^2 \cdot 5$	7	15 cm
8	$6\frac{1}{6}$	8	$-3 = a$	8	15 teaspoons
9	1,760 yards	9	2, 3, 5, 7, 11, 13	9	36 feet
10	18	10	36 pieces	10	0.80, 80%
11	$x = -37$	11	$x = -195$	11	$\frac{2x}{7} = 40$
12	$2 \cdot 7 \cdot a \cdot a \cdot a \cdot b \cdot b$	12	0.39	12	$2a^2bc$
13	$1 = m$	13	25	13	$x = 7$
14	20	14	$18x^2y^3z^2$	14	-56
15	$A = bh$	15	$x = 210$	15	13
16	100°C	16		16	90%
17	$b = -17$	17	nine thousand three hundred fifteen ten thousandths	17	-26
18	$2ab$	18	$x = -3$	18	$b = -54$
19	$1\frac{1}{10}$	19	$\frac{2}{3}$	19	$x = 80$
20	a) Q b) M	20	42 cm³	20	4,044

	Lesson #22		Lesson #23		Lesson #24
1	$x = 17$	1	-133	1	$-1 = x$
2	$24x^3y^4z^2$	2	17	2	$10x^2y^2z$
3	29	3	$<$	3	50.24 cm
4	378 in.²	4	16 cups	4	$x = -105$
5	67	5	$-2 = a$	5	324 inches
6	$x = 208$	6	6	6	$x = 2$
7	83	7	212°F	7	$x = 3$
8	54°	8	1.55	8	$12a - 12b + 14$
9	26,400 feet	9	$4x^3yz$	9	36
10	$b = -9$	10	63 m²	10	-81
11	$\frac{9}{20}$	11	15%	11	0.7134
12	$2 \cdot 5 \cdot a \cdot a \cdot a \cdot b$	12	$\frac{3}{4}$	12	$\frac{2xz}{5}$
13	76.214	13	$x - 14$	13	15
14	$5\frac{2}{3}$	14	3.03	14	$6\frac{4}{9}$
15	20	15	$3\frac{1}{5}$	15	78 m²
16	$x = 108$	16	$b = 95$	16	$x = 6$
17	$b = 16$	17	$5 \cdot 5 \cdot a \cdot a \cdot b \cdot c$	17	9
18	Wednesday	18	$x = 210$	18	60
19	74°	19	$x = 72$	19	-79
20	14°	20	0.0000072	20	A(−5, −5) B(4, 1) C(0, −2)

	Lesson #25		Lesson #26		Lesson #27
1	-19	**1**	$12.50	**1**	$y = 10$
2	$48a^2b^2c^5$	**2**	$20x - 33y + 79$	**2**	$8x = 16$
3	$x = 95$	**3**	$-3 = x$	**3**	$\frac{1}{32}$
4	95 pounds	**4**	21	**4**	$\begin{pmatrix} 1 & -3 \\ 5 & 1 \end{pmatrix}$
5	$\frac{xy}{5z}$	**5**	$x = 35$	**5**	36 mm²
6	$\frac{10}{2x}$	**6**	12 quarts	**6**	30
7	$x = 16$	**7**	$b = 39$	**7**	$11\frac{17}{18}$
8	$9\frac{17}{20}$	**8**	$2\frac{2}{3}$	**8**	0.14
9	95%	**9**	$4xyz^2$	**9**	3 yards
10	decagon	**10**	$x = 52$	**10**	$\frac{x^2y}{2}$
11	$x = 120$	**11**	16	**11**	$x = 65$
12	$\frac{3}{4}$	**12**	$20\frac{1}{5}$	**12**	$A = bh$
13	$-8 = m$	**13**	78.5 cm²	**13**	$36a^3b^2c$
14	-210	**14**		**14**	1
15	$0.48;\ \frac{12}{25}$	**15**	$\frac{2ab^3c}{3}$	**15**	$x = -25$
16	67	**16**	113°	**16**	41,725
17	5	**17**	0.00020	**17**	112 watermelons
18	$x = -3$	**18**	72 inches	**18**	3.25
19	5,280 yards	**19**	$x = -3$	**19**	$x = 1$
20	15°, 45°, 80°	**20**	D(0, 5) F(5, 0)	**20**	supplementary

	Lesson #28		Lesson #29		Lesson #30
1	$28a - 12b + 14$	1	10,560 feet	1	$\frac{3}{4}$
2	(parallelogram)	2	0.65	2	$a = -12$
3	$x = 7$	3	64	3	$\frac{x}{6}$
4	64	4	1	4	$x = -77$
5	150 ft^2	5	$a = 45$	5	$\frac{1}{3}$
6	$\frac{1}{2}$	6	241	6	$\frac{2x^2y^2z}{3}$
7	32	7	$x = 48$	7	$-10 = x$
8	$x = 6$	8	$\frac{3x}{2} - 4$	8	90 m^2
9	-52	9	$\begin{pmatrix} -9 & 7 & -3 \\ -3 & 10 & -13 \end{pmatrix}$	9	-96
10	$x = 66$	10	$\frac{x^2z}{3}$	10	180°
11	$5\frac{5}{7}$	11	$x = -16$	11	$6a - 29b - 8$
12	$5xy^2z^3$	12	0.12	12	$56x^2y^3z^2$
13	24	13	6.4	13	2, 3, 5, 7
14	<	14	(two double-headed arrows)	14	$x = 72$
15	$a = -90$	15	$30x^2y^3z$	15	250,000,000
16	41	16	32°F	16	36
17	17,461,599	17	$x = 3$	17	three and four hundred seventy nine thousandths
18	$\frac{ab}{3}$	18	12	18	$x = 5$
19	0.60, 60%	19	11	19	42%, $\frac{21}{50}$
20	a) Z b) Y	20	$46° = x$	20	J(1, 4) K(−3, 0)